Noise and Vibration Control

Second Edition

IISc Lecture Notes Series

ISSN: 2010-2402

Editor-in-Chief: Diptiman Sen

World Scientific Publishing Company Singapore and Indian Institute of Science (IISc), Bangalore, will co-publish a series of prestigious lectures delivered during IISc's centenary year (2008–09), and a series of textbooks and monographs, by prominent scientists and engineers from IISc and other institutions.

This pioneering collaboration will contribute significantly in disseminating current Indian scientific advancement worldwide. In addition, the collaboration also proposes to bring the best scientific ideas and thoughts across the world in areas of priority to India through specially designed India editions.

The "IISc Lecture Notes Series" is based on postgraduate courses developed and taught in IISc, and in other major postgraduate programs in India. These are class tested, compact, can be used directly as lectures.

Published:

For the complete list of volumes in this series, please visit
www.worldscientific.com/series/iislns

IISc Lecture Notes Series

Noise and Vibration Control

Second Edition

M L Munjal

Indian Institute of Science, India

B Venkatesham

Indian Institute of Technology Hyderabad, India

IISc
Press

World Scientific

NEW JERSEY · LONDON · SINGAPORE · BEIJING · SHANGHAI · HONG KONG · TAIPEI · CHENNAI · TOKYO

Published by

World Scientific Publishing Co. Pte. Ltd.

5 Toh Tuck Link, Singapore 596224

USA office: 27 Warren Street, Suite 401-402, Hackensack, NJ 07601

UK office: 57 Shelton Street, Covent Garden, London WC2H 9HE

Library of Congress Cataloging-in-Publication Data
Names: Munjal, M. L. (Manohar Lal), 1945–   author. | Venkatesham, B., author.
Title: Noise and vibration control / M L Munjal, Indian Institute of Science, India,
    B Venkatesham, Indian Institute of Technology Hyderabad, India.
Description: Second edition. | Singapore : IISc Press ; Hackensack, New Jersey :
    World Scientific, [2025] | Series: IISc lecture notes series, 2010-2402 ; vol. 8 |
    Includes bibliographical references and index.
Identifiers: LCCN 2024002524 | ISBN 9789811283147 (hardcover) |
    ISBN 9789811283154 (ebook for institutions) | ISBN 9789811283161 (ebook for individuals)
Subjects: LCSH: Noise control. | Vibration.
Classification: LCC TD892 .M85 2025 | DDC 620.3/7--dc23/eng/20240314
LC record available at https://lccn.loc.gov/2024002524

British Library Cataloguing-in-Publication Data
A catalogue record for this book is available from the British Library.

For any available supplementary material, please visit
https://www.worldscientific.com/worldscibooks/10.1142/13589#t=suppl

Desk Editors: Nandha Kumar/Steven Patt

Typeset by Stallion Press
Email: enquiries@stallionpress.com

Printed in Singapore

In pursuit of Quietness

Series Preface

World Scientific Publishing Company - Indian Institute of Science Collaboration

IISc Press and WSPC are co-publishing books authored by world renowned scientists and engineers. This collaboration, started in 2008 during IISc's centenary year under a Memorandum of Understanding between IISc and WSPC, has resulted in the establishment of three series: IISc Centenary Lecture Series (ICLS), IISc Research Monographs Series (IRMS), and IISc Lecture Notes Series (ILNS).

This pioneering collaboration will contribute significantly in disseminating current Indian scientific advancement worldwide.

The **"IISc Centenary Lecture Series"** will comprise lectures by designated Centenary Lecturers - eminent teachers and researchers from all over the world.

The **"IISc Research Monographs Series"** will comprise state-of-the-art monographs written by experts in specific areas. They will include, but not limited to, the authors' own research work.

The **"IISc Lecture Notes Series"** will consist of books that are reasonably self-contained and can be used either as textbooks or for self-study at the postgraduate level in science and engineering. The books will be based on material that has been class-tested for most part.

Diptiman Sen, Editor-in-Chief (diptiman@iisc.ac.in)

Preface to the Second Edition

The original edition of *Noise and Vibration Control*, published about a decade ago, was generally well received by practicing noise control engineers and designers of quieter layouts as well as senior undergraduates in mechanical engineering. This second edition is intended as an update.

To make the book more useful for the intended readership, academic as well as professional, three new chapters have been added: one on acoustic measurements (Chapter 2), one on sound transmission through media (Chapter 5), and one on computational acoustics (Chapter 9). These additions are largely the contribution of the second author (Prof. B. Venkatesham) who has been using the original edition of the book for teaching the graduate as well as undergraduate students of the Indian Institute of Technology Hyderabad (IIT Hyderabad), offering short-term courses to instructors, industrial professionals, and graduate students from other academic institutes under the Continuing Education Program, and also for carrying out several industrial consultancy projects on industrial, environmental and automotive noise control. While preparing the textbook content for this second edition, he has also drawn on his decade-long industrial experience prior to his academic career.

While drafting the additional Chapters 2, 5 and 9, he relied primarily on the joint publications of his previous as well as current graduate and research students. He wishes to particularly thank Drs. Nagaraja Jade, Tapan Mahanta, Deepak Akiwate,

Veerababu and Sivateja, and his graduate students Yoganand and Hariharasudhan.

He also wants to thank his wife Anita alias Likhita and daughters Hamsini and Sahasra for their encouragement and understanding of the time commitment required to complete the book outside of the normal working hours and during weekends.

He is grateful to the current and prior Heads of the Department (HoD) of Mechanical and Aerospace Engineering and the current Director of IIT Hyderabad for providing facilities and promoting research, teaching, consultancy, and book writing. Prof. U. B. Desai, founding director of IIT Hyderabad, deserves a special mention for his outstanding leadership and encouragement during the author's early academic career at IIT Hyderabad.

M. L. Munjal
Bangalore
March 2024

B. Venkatesham
Hyderabad
March 2024

Preface to the First Edition

Noise is defined as unwanted sound. Excessive or persistent noise may cause annoyance, speech interference and hearing damage. In a working environment, noise may lead to several physiological disorders like high blood pressure, heart problems, headache, etc. Noise is also known to cause accidents at the workplace and loss of efficiency and productivity.

Vibration is caused by unbalanced inertial forces and moments. Resonant vibrations may lead to fatigue failures. Flexural vibrations of the exposed surfaces of a machine radiate audible noise, and in fact represent one of the primary sources of noise. Excessive vibration and noise characterize all rotating, reciprocating and flow machinery. This makes automobiles, aeroplanes, thermal power stations, etc. excessively noisy. Thus, the problems of noise and vibration are ubiquitous, cutting across all disciplines of engineering. This book deals primarily with industrial and automotive noise, its measurement and control. The control of noise from vibrating bodies at the source involves control of vibration. Therefore, two of the six chapters deal with vibration, its measurement and control.

It is now well understood that a quieter machine is in every way a better machine. Lesser vibration ensures manufacturing closer tolerances, lesser wear and tear, and longer fatigue life. Hence, a quieter machine is more cost-effective in the long run. Designing for quietness is known to be most cost-effective. Noise control of existing machinery, while often necessary, calls for stoppage of the machinery and excessive retrofit costs.

The All-India Council for Technical Education (AICTE) has listed a course on "Noise and Vibration Control" as a possible elective course for senior undergraduates of the engineering colleges in the country. Such a course would need an appropriate textbook. Hence, this presentation.

The author has been teaching this course at the graduate level at the Department of Mechanical Engineering of the Indian Institute of Science (IISc) for over three decades. With a good number of solved as well as unsolved exercises, the present textbook lays stress on design methodologies, applications and exercises. Analytical derivations and techniques are eschewed. Nevertheless, references are provided at the end of each chapter for further study.

This textbook stresses on physical concepts and the application thereof to practical problems. The author's four decades experience in teaching, research and industrial consultancy is reflected in the choice of the solved examples and unsolved problems. The book targets senior undergraduate mechanical engineering students as well as designers of industrial machinery and layouts. It can readily be used for self-study by practicing designers and engineers. This is why mathematical derivations have been avoided. The illustrations, tables and empirical formulae have been offered for ready reference.

As Chairman and Member Secretary of the Steering Committee of the Facility for Research in Technical Acoustics (FRITA), Professor D. V. Singh and Mr. S. S. Kohli have played an important role in conceptualizing and supporting this book writing project.

This book has been influenced substantially by Professor Colin H. Hansen whose book *Engineering Noise Control* I have been following in my graduate course at IISc, and Mr. D. N. Raju with whom I have been collaborating on many of my consultancy projects — in pursuit of quietness. I wish to acknowledge the personal inspiration of Professors Malcolm J. Crocker, M. V. Narasimhan, B. V. A. Rao, S. Narayanan, B. C. Nakra, A. K. Mallik and D. N. Manik, among others.

I have drawn heavily from the joint publications of my past as well as present graduate students and research students. My sincere thanks to all of them, particularly, Dr. Prakash T. Thawani and Professor Mohan D. Rao.

I thank Professor R. Narasimhan, Chairman of the Department of Mechanical Engineering, Indian Institute of Science for providing all facilities as well as a conducive environment for research, teaching, consultancy and book writing.

I wish to thank my wife Vandana alias Bhuvnesh for bearing with me during long evenings and weekends that were needed to complete the book.

This textbook has been catalyzed and supported by the Department of Science and Technology (DST), under its Utilization of Scientific Expertise of Retired Scientists (USERS) scheme.

M. L. Munjal
Bangalore
April 2013

Contents

Chapter 1

Introduction to Acoustics

Sound is a longitudinal wave in air, and wave is a traveling disturbance. Mass and elasticity of the air medium are primary characteristics for a wave to travel from the source to the receiver. A wave is characterized by two state variables, namely, pressure and particle velocity. These represent perturbations on the static ambient pressure and the mean flow velocity of wind, respectively. The perturbations depend on time as well as space or distance.

Noise is unwanted sound. It may be unwanted or undesirable because of its loudness or frequency characteristics. Excessive or prolonged exposure to noise may lead to several physiological effects like annoyance, headache, increase in blood pressure, loss of concentration, speech interference, loss of working efficiency, or even accidents in the workplace. Persistent exposure of a worker to loud noise in the workplace may raise his/her threshold of hearing.

The study of generation, propagation and reception of audible sound constitutes the science of Acoustics. There are several branches of acoustics, namely, architectural acoustics, electroacoustics, musical acoustics, underwater acoustics, ultrasonics, physical acoustics, etc. The field of industrial noise, automotive noise and environmental noise constitutes engineering acoustics or technical acoustics. This in turn comprises sub-areas like duct acoustics, vibro-acoustics, computational acoustics, etc.

The speed at which the longitudinal disturbances travel in air is called sound speed, c. It depends on the ambient temperature, pressure and density as follows:

$$c = (\gamma R T)^{1/2} = (\gamma p_0 / \rho_0)^{1/2} \qquad (1.1)$$

Here, γ is the ratio of specific heats C_p and C_v, R is gas constant, p_0 is static ambient pressure, p_0 is mass density, and T is the absolute temperature of the medium. For air at standard pressure ($\gamma = 1.4$, $R = 287.05 \, \text{J/(kg·K)}$, $p_0 = 1.013 \times 10^5 \, \text{Pa}$), it can easily be seen that

$$c \simeq 20.05(T)^{1/2} \qquad (1.2)$$

where T is the absolute temperature in Kelvin.

Symbol T is used for the time period of harmonic disturbances as well. It is related to frequency f as follows:

$$T = 1/f \quad \text{or} \quad f = 1/T \qquad (1.3)$$

Frequency f is measured in Hertz (Hz) or cycles per second. Wavelength λ of moving disturbances, of frequency f, is given by

$$\lambda = c/f \qquad (1.4)$$

where c denotes speed of sound.

1.1 Plane Wave Propagation

Plane waves moving inside a waveguide (a duct with rigid walls) are called one-dimensional (1D) waves. These are characterized by the following 1D wave equation [1]:

$$\frac{\partial^2 p}{\partial t^2} - c^2 \frac{\partial^2 p}{\partial z^2} = 0 \qquad (1.5)$$

where p, z and t are acoustic pressure, coordinate along direction of wave propagation and time, respectively.

For harmonic waves, the time dependence is given by $e^{j\omega t}$ or $\cos(\omega t)$ or $\sin(\omega t)$, where $\omega = 2\pi f$ is the circular frequency in rad/s.

General solution of Eq. (1.5) may be written as

$$p(z,t) = (Ae^{-jkz} + Be^{jkz})e^{j\omega t} \tag{1.6}$$

or as

$$p(z,t) = Ae^{j\omega(t-z/c)} + Be^{j\omega(t+z/c)} \tag{1.7}$$

where $k = \omega/c = 2\pi/\lambda$ is called the wave number.

It can easily be seen that A is amplitude of the forward progressive wave and B is amplitude of the reflected or rearward progressive wave. Algebraic sum of the two progressive waves moving in opposite directions is called a standing wave. Thus, Eq. (1.6) represents acoustic pressure of a 1D standing wave. The corresponding equation for particle velocity is given by

$$u(z,t) = \frac{1}{\rho_0 c}(Ae^{-jkz} - Be^{jkz})e^{j\omega t} \tag{1.8}$$

$\rho_0 c$, product of the mean density and sound speed, represents the characteristic impedance of the medium.

For an ambient temperature of 25°C and the standard atmospheric pressure (corresponding to the mean sea level), we have

$$p_0 = 1.013 \times 10^5 \,\text{Pa}, \quad T = 298 \,\text{K}, \quad \rho_0 = 1.184 \,\text{kg/m}^3,$$

$$c = 346 \,\text{m/s}, \quad \rho_0 c = 410 \,\text{kg/(m}^2\text{s)}$$

1D wave occurs primarily in the exhaust and tailpipe of automotive engines and reciprocating compressors. These waves are characterized by a plane wavefront normal to the axis of the pipe or tube, and therefore they are called plane waves.

The forward wave is generated by the source and the rearward wave is the result of reflection from the passive termination downstream. In particular, $B/A = R$ is called the Reflection Coefficient, and may be determined from the termination impedance[1]. In particular, $R = 0$ for anechoic termination, 1 for rigid (closed) termination, and 0 for expansion into vacuum. In general, R is a function of frequency.

In view of the plane wave character of the 1D waves, the acoustic power flux W of a plane progressive wave may be written as

$$W = IS = \langle pu \rangle S = \langle p\nu \rangle, \quad \nu = S\,u \tag{1.9}$$

where I is Sound Intensity defined as power per unit area in a direction normal to the wavefront (in the axial direction for plane waves), S is area of cross-section of the pipe, and ν is called the Volume Velocity.

Thus, the acoustic power flux associated with the incident progressive wave and the reflected progressive wave are given by

$$W_i = \frac{|A|^2 S}{2\rho_0 c} \tag{1.10}$$

$$W_r = \frac{|B|^2 S}{2\rho_0 c} \tag{1.11}$$

Note that the factor of 2 in the denominator is due to the mean square values required in the power calculations. Thus, the net power associated with a standing wave is given by

$$W \equiv W_i - W_r = \frac{|A|^2 - |B|^2}{2(\rho_0 c / S)} = \frac{|A|^2 - |B|^2}{2Y} \tag{1.12}$$

where $Y = \rho_0 c / S$ is the Characteristic Impedance of plane waves, defined as ratio of acoustic pressure and volume velocity of a plane progressive wave along a tube of area of cross-section S.

1.2 Spherical Wave Propagation

Wave propagation in free space is characterized by the following 3D wave equation [1]:

$$\left[\frac{\partial^2}{\partial t^2} - c^2 \nabla^2 \right] p = 0 \tag{1.13}$$

where ∇^2 is the Laplacian. In terms of spherical polar coordinates, neglecting angular dependence for spherical waves, Eq. (1.13) can be written as

$$\frac{\partial^2 (rp)}{\partial t^2} - c^2 \frac{\partial^2 (rp)}{\partial r^2} = 0 \tag{1.14}$$

Comparison of Eqs. (1.5) and (1.14) suggests the following solution for spherical waves:

$$p(r,t) = \frac{1}{r}\{Ae^{-jkr} + Be^{jkr}\}e^{j\omega t} \tag{1.15}$$

Substituting it into the momentum equation

$$\rho_0 \frac{\partial u}{\partial t} = -\frac{\partial p}{\partial r} \tag{1.16}$$

yields

$$u(r,t) = \frac{j}{\omega \rho_0 r} \left\{ -\left(jk + \frac{1}{r}\right) Ae^{-jkr} + \left(jk - \frac{1}{r}\right) Be^{jkr} \right\} e^{j\omega t} \tag{1.17}$$

Here r is the radial distance between the receiver and a point source. It may again be noted that the first component of Eqs. (1.15) and (1.17) represent the spherically outgoing or diverging wave and the second one represents the incoming or converging spherical wave. In practice, the second component is hypothetical; in all practical problems dealing with noise radiation from vibrating bodies one deals with the diverging wave only. The ratio of pressure and particle velocity for the diverging progressive wave may be seen to be

$$\frac{p(r,t)}{u(r,t)} = \frac{\omega \rho_0}{k - j/r} = \frac{\rho_0 c}{1 - \frac{j}{kr}} = \rho_0 c \frac{jkr}{1 + jkr} \tag{1.18a}$$

It may be observed that unlike for plane progressive waves, this ratio is a function of distance r. This indicates that for a spherical diverging wave, pressure and velocity are not in phase. However, when the Helmholtz number kr tends to infinity (or is much larger than unity), then this ratio tends to $\rho_0 c$. Physically, it implies that in the far field a spherical diverging wave becomes or behaves as a plane progressive wave. This also indicates that the microphone of the sound level meter should not be near the vibrating surface; it should be in the far field.

In the far field, Helmholtz number is much larger than unity $(kr \gg 1)$ and then Eqs. (1.15), (1.17) and (1.18a) reduce to,

$$p(r,t) = \frac{A}{r} e^{-jkr} e^{j\omega t}, \quad u(r,t) = p(r,t)/(\rho_0 c) \qquad (1.18b)$$

and therefore, sound intensity and total power are given by:

$$I(r) = \frac{\text{Re}(p(r)u^*(r))}{2} = \frac{|p(r)|^2}{2\rho_0 c} = \frac{|A|^2}{2\rho_0 cr^2} = \frac{W}{4\pi r^2} \qquad (1.18c)$$

where $4\pi r^2$ is the surface area of a hypothetical sphere of radius r over which the total power W is divided equally to yield intensity $I(r)$.

Example 1.1. A bubble-like sphere of 5 mm radius is pulsating harmonically at a frequency of 1000 Hz with amplitude of radial displacement 1 mm in air at mean sea level and 25°C. Evaluate:

(a) amplitude of the radial velocity of the sphere surface,
(b) amplitude of the acoustic pressure and particle velocity at a radial distance of 1 m.

Solution.

(a) For harmonic radial motion, radial velocity u equals ω times the radial displacement ξ, where $\omega = 2\pi f$. Thus

$$\omega = 2\pi \times 1000 = 6283.2 \, \text{rad/s}$$

$$|u| = \omega|\xi| = 6283.2 \times \frac{1}{1000} = 6.283 \, \text{m/s}$$

(b) Wave number, $k = \frac{\omega}{c} = \frac{6283.2}{346} = 18.16 \, \text{m}^{-1}$

Distance, $r = 1 \, \text{m}$ (given)

Helmholtz number, $kr = 18.16$ at $r = 1 \, \text{m}$, and 0.091 at $r = 0.005 \, \text{m}$ (i.e., on the surface). As per Eq. (1.17), for a diverging spherical wave in free field, $B = 0$, and

$$|u|_{\text{surface}} = \frac{(1 + k^2 r_0^2)^{1/2}}{\omega \rho_0 r_0^2} A$$

or

$$A = \frac{|u|_{\text{surface}} \cdot \omega \rho_0 r_0^2}{(1 + k^2 r_0^2)^{1/2}} = \frac{6.283 \times 6283.2 \times 1.184 \times (0.005)^2}{\{1 + (0.091)^2\}^{1/2}}$$

$$= \frac{1.1685}{1.004} = 1.1638$$

Substituting this value of A in Eq. (1.17) for $r = 1$m yields

$$|u|_{r=1\,\text{m}} = A \left. \frac{(1 + k^2 r^2)^{1/2}}{\omega \rho_0 r^2} \right]_{r=1\,\text{m}}$$

$$= 1.1638 \times \frac{\{1 + (18.16)^2\}^{1/2}}{6283.2 \times 1.184 \times (1)^2} = 2.845 \times 10^{-3}\,\text{m/s}$$

Now, use of Eq. (1.15) yields

$$|p|_{r=1\,\text{m}} = \frac{A}{r} = \frac{1.1638}{1} = 1.164\,\text{Pa}$$

Incidentally, sound pressure amplitude at 1 m may also be obtained by means of Eq. (1.18a):

$$|p|_{r=1\,\text{m}} = |u|_{r=1\,\dot{\text{m}}} \, \rho_0 c \frac{18.16}{\{1 + (18.16)^2\}^{1/2}}$$

$$\simeq |u|_{r=1\,\dot{\text{m}}} \, \rho_0 c$$

$$= 2.845 \times 10^{-3} \times 410 = 1.166\,\text{Pa}$$

It is worth noting that

$$\frac{p}{u} = \rho_0 c \times jkr \text{ at the surface, where } kr \ll 1,$$

$$\frac{p}{u} = \rho_0 c \text{ in the farfield, where } kr \gg 1,$$

Thus, at the surface (or in the nearfield), sound pressure leads radial velocity by $90°$ (because $j = e^{j\pi/2}$), whereas in the farfield sound pressure is in phase with particle velocity.

1.3 Wave Characteristics

Acoustic waves are mechanical longitudinal waves propagating through a medium, typically air, but can also travel through liquids and solids. The acoustic wave characteristics are like other wave types: optical, electromagnetic, seismic, and Rayleigh. The typical wave characteristics are frequency, wavelength, wave speed, phase, attenuation, interference, reflection, refraction, scattering, and diffraction. Equation (1.4) gives the relationship of frequency, wavelength, and speed of sound.

Acoustic wave reflection, scattering, and diffraction are important phenomena that occur when sound waves encounter obstacles, boundaries, or other variations in the medium through which they are traveling. These phenomena are fundamental in understanding how sound propagates and interacts with its environment. Understanding these characteristics is helpful in applications to building acoustics, sonar technology, audio engineering, and ultrasound imaging.

Reflection is the process by which sound waves bounce off a surface or boundary and return to the medium from which they originated. When an acoustic wave encounters a reflective surface, it undergoes a change in direction and continues to propagate through the medium. The angle of incidence (θ_i) is equal to the angle of reflection (θ_r), as shown in Fig. 1.1. Echoes are created by reflection phenomenon.

Scattering occurs when sound waves encounter small or irregular objects or variations in the medium that cause the waves to be redirected in various directions. During scattering, the sound waves

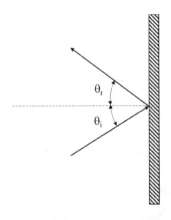

Fig. 1.1 Schematic diagram of reflection.

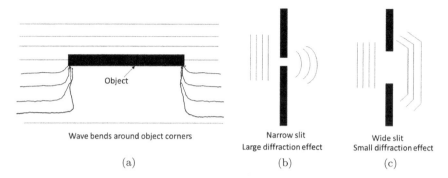

Fig. 1.2   Wave diffraction phenomena in various configurations.

are dispersed in multiple directions and create a diffuse field rather than simply reflecting off a surface.

Diffraction is the bending or spreading of sound waves as they encounter an obstacle or an aperture that is on the order of or smaller than, the wavelength of the sound. The degree of diffraction is influenced by the wavelength of the sound and the size of the obstacle or aperture. Smaller wavelengths and larger openings lead to more pronounced diffraction, as shown in Fig. 1.2. Diffraction is responsible for the ability of sound to bend around obstacles, allowing us to hear sounds even when the source is not directly visible. This wave property is used in the development of noise barriers.

Refraction is a phenomenon that occurs when sound waves change their direction and speed as they travel through a medium with varying acoustic properties, or transition from one medium to another with different acoustic properties. The change in direction and speed of the wave can be related by Snell's Law. It can be written as

$$\frac{\sin\theta_1}{\sin\theta_2} = \frac{n_1}{n_2} \tag{1.19}$$

where θ_1, θ_2 are the angles of incidence and refraction, respectively, and n_1, n_2 are the refractive indices of the two media.

Refraction typically occurs at the boundary or interface between two different media, e.g., air and water, or glass and air. Sound wave

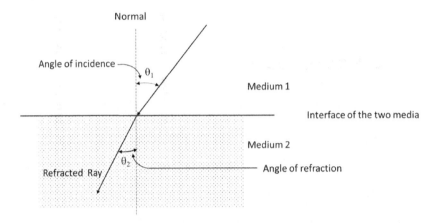

Fig. 1.3 Schematic diagram of refraction (bends due to change in medium).

bends towards the surface normal when the wave passes from low-density to high-density medium, as shown in Fig. 1.3. The wave moves away from the surface normal when the wave propagates from high-density medium to low-density medium. Refraction phenomena have applications in environmental acoustics and underwater acoustics.

The Doppler effect elucidates alterations in the perceived frequency or wavelength of a sound wave based on the motion of the source in relation to the receiver. When both the source and receiver are stationary, the apparent frequency (f') at the receiver is the same as the frequency produced by the source. However, as the source approaches the receiver, the apparent frequency increases, decreasing when the source moves away, as shown in Fig. 1.4.

If the source travels faster than sonic speed, the source reaches the receiver before the sound. Sound and source reach the receiver at the same time for the sonic speed of the source.

Example 1.2. A train approaches the railway platform at the speed of 54 kmph and driver starts the honking for safety reasons while the receiver is standing on the platform. The honking frequency is 400 Hz. Calculate the apparent frequency in approach and leaving the platform. The speed of sound is 340 m/s.

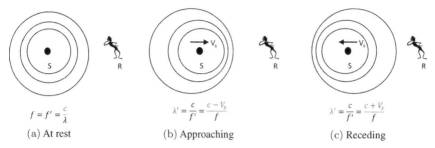

$$f = f' = \frac{c}{\lambda}$$

(a) At rest

$$\lambda' = \frac{c}{f'} = \frac{c - V_s}{f}$$

(b) Approaching

$$\lambda' = \frac{c}{f'} = \frac{c + V_s}{f}$$

(c) Receding

Fig. 1.4 Doppler effect and change in apparent frequency for source movement.

Solution. Given

$$V_s = 54\,\text{kmph}, \quad f = 400\,\text{Hz}, \quad c = 340\,\text{m/s}$$

Converting train velocity from kmph to m/s

$$V_s = 54 * 1000/3600 = 15\,\text{m/s}$$

Apparent frequency while approaching the platform (Fig. 1.4(b)):

$$f' = \frac{cf}{c - V_s}$$
$$f' = \frac{340 * 400}{340 - 15} = 418.5\,\text{Hz}$$

Apparent frequency while leaving the platform (Fig. 1.4(c)):

$$f' = \frac{cf}{c + V_s}$$
$$f' = \frac{340 * 400}{340 + 15} = 383\,\text{Hz}$$

1.4 Decibel Level

Human ear is a fantastic transducer. It can pick up pressure fluctuations of the order of 10^{-5} Pa to 10^3 Pa; that is, it has a dynamic range of 10^8. Therefore, a linear unit of measurement is ruled out. Instead, universally a logarithmic unit of decibels has been adopted for measurements of Sound Pressure Level, Intensity Level and Power Level. These are defined as follows [1–3]:

$$SPL \equiv L_p = 10\log \frac{p_{rms}^2}{p_{th}^2} = 20\log\left(\frac{p_{rms}}{2 \times 10^{-5}}\right), \text{ dB} \quad (1.20)$$

$$IL \equiv L_I = 10\log \frac{I}{I_{ref}} = 10\log\left(\frac{I}{10^{-12}}\right), \text{ dB} \quad (1.21)$$

$$SWL \equiv L_w = 10\log \frac{W}{W_{ref}} = 10\log\left(\frac{W}{10^{-12}}\right), \text{ dB} \quad (1.22)$$

where log denotes log to the base 10, and $p_{th} = 2 \times 10^{-5}$ Pa represents the threshold of hearing. This standard quantity represents the root-mean-square pressure of the faintest sound of 1000 Hz frequency that a normal human ear can just pick up. The corresponding value of the reference intensity represents,

$$I_{ref} = \frac{p_{th}^2}{\rho_0 c} \approx \frac{(2 \times 10^{-5})^2}{400} = 10^{-12} \text{ W/m}^2 \quad (1.23)$$

Similarly, $W_{ref} = 10^{-12}$ W.

Making use of Eqs. (1.15) and (1.17) in Eq. (1.21) indicates that in the far field the acoustical intensity is inversely proportional to square of the radial distance r. This inverse square law when interpreted in logarithmic units becomes,

$$L_I(2r) - L_I(r) = 10\log \frac{r^2}{(2r)^2} = -6\,\text{dB} \quad (1.24)$$

This indicates that in the far field, the sound pressure level or sound intensity level would decrease by 6 dB when the measurement distance from the source is doubled.

For spherically diverging waves the sound pressure level at a distance r from a point source is related to the total sound power level as follows [2]:

$$L_p(r) = L_W + 10\log\left(\frac{Q}{4\pi r^2}\right), \text{ dB} \quad (1.25)$$

where Q is the locational directivity factor, given by [2, 3]

$$Q = 2^{n_s} \tag{1.26}$$

Here, n_s is the number of surfaces touching at the source. Thus,

$n_s = 0$ for a source in midair (or free space),
1 for a source lying on the floor,
2 for a source located on the edge of two surfaces, and
3 for a source located in a corner (where three surfaces meet).

Specifically, for a source located on the floor in the open, Eq. (1.25) yields,

$$L_p(r) = L_W - 10 \log(2\pi r^2), \text{ dB} \tag{1.27}$$

1.5 Frequency Analysis

The human ear responds to sounds in the frequency range of 20–20,000 Hz (20 kHz), although the human speech range is 125–8000 Hz. Precisely, male speech lies between 125 Hz and 4000 Hz, and female speech is one octave higher, that is, 250–8000 Hz.

The audible frequency range is divided into octave and 1/3-octave bands. For an octave band,

$$f_u/f_l = 2 \quad \text{and} \quad f_m = (f_u \cdot f_l)^{1/2} \tag{1.28}$$

so that

$$f_l = \frac{f_m}{2^{1/2}} = 0.707 f_m \quad \text{and} \quad f_u = f_m \cdot 2^{1/2} = 1.414 f_m \tag{1.29}$$

Similarly, for a one-third octave band,

$$f_u/f_l = 2^{1/3} \quad \text{and} \quad f_m = (f_u \cdot f_l)^{1/2} \tag{1.30}$$

so that

$$f_l = \frac{f_m}{2^{1/6}} = 0.891 f_m \quad \text{and} \quad f_u = f_m \cdot 2^{1/6} = 1.1225 f_m \tag{1.31}$$

In Eqs. (1.28) to (1.31), subscripts l, u and m denote lower, upper and mean, respectively. It may be noted that three contiguous

1/3-octave bands would have the combined frequency range of the octave band centered at the centre frequency of the middle 1/3-octave band.

1000 Hz has been recognized internationally as the standard reference frequency, and the mid frequencies of all octave bands and 1/3-octave bands have been fixed around this frequency. Table 1.1 gives a comparison of different octave and 1/3-octave bands.

Incidentally, the standard frequency of 1000 Hz happens to be the geometric mean of the human speech frequency range; that is,

$$1000 = (125 * 8000)^{1/2} \tag{1.32}$$

It may also be noted that,

$$2^{1/3} = 1.26 \quad \text{and} \quad 10^{1/10} = 1.259 \tag{1.33}$$

Thus, for practical purposes, $2^{1/3} = 10^{1/10}$. That is why working either way around 1000, 100, 200, 500, 2000, 5000, and 10,000 Hz represent the mean frequencies of the respective 1/3-octave bands. It may also be noted that the center frequencies indicated in the second and fifth columns of Table 1.1 are internationally recognized nominal frequencies and may not be precise.

It may also be noted that the octave band and 1/3-octave band filters are constant percentage bandwidth filters. The percentage bandwidth of an n-octave filter may be written as

$$bw_n \equiv \frac{f_u - f_l}{f_m} \times 100 = [2^{n/2} - 2^{-(n/2)}] \times 100 \tag{1.34}$$

Thus, for an octave filter ($n = 1$), bandwidth is 70.7% and for a one-third octave filter ($n = 1/3$), the bandwidth is 23.16% of the mean or centre frequency of the particular filter.

Power spectral density represents acoustic power per unit frequency as a function of frequency. Power in a band of frequencies represents area under the curve within this band. For example, for a flat (constant power) spectrum, the power in a frequency band would be proportional to the bandwidth in Hertz. As the bandwidth of an octave band doubles from one band to the next, the sound

Table 1.1.   Bandwidth and geometric mean frequency of standard octave and 1/3-octave bands [12].

1 Octave			1/3 Octave		
Lower cutoff frequency (Hz)	Center frequency (Hz)	Upper cutoff frequency	Lower cutoff frequency (Hz)	Center frequency (Hz)	Upper cutoff frequency
			22.4	25	28.2
22	31.5	44	28.2	31.5	35.5
			35.5	40	44.7
			44.7	50	56.2
44	63	88	56.2	63	70.8
			70.8	80	89.1
			89.1	100	112
88	125	177	112	125	141
			141	160	178
			178	200	224
177	250	355	224	250	282
			282	315	355
			355	400	447
355	500	710	447	500	562
			562	630	708
			708	800	891
710	1000	1420	891	1000	1122
			1122	1250	1413
			1413	1600	1778
1420	2000	2840	1778	2000	2239
			2239	2500	2818
			2818	3150	3548
2840	4000	5680	3548	4000	4467
			4467	5000	5623
			5623	6300	7079
5680	8000	11,360	7079	8000	8913
			8913	10,000	11,220
			11,220	12,500	14,130
11,360	16,000	22,720	14,130	16,000	17,780
			17,780	20,000	22,390

pressure level or power level would increase by $10 \log 2 = 3 \, \text{dB}$. Similarly, the SPL or SWL of a 1/3-octave band would increase by $10 \log 2^{1/3} = 1 \, \text{dB}$, as we move from one band to the next. As a corollary of the phenomenon, SPL in an octave band would be equal to the logarithmic sum of the SPLs of the three contiguous 1/3-octave bands constituting the octave band.

1.6 Weighted Sound Pressure Level

The human ear responds differently to sounds of different frequencies. Extensive audiological surveys have resulted in weighting factors for different purposes. Originally, A-weighting was for sound levels below 55 dB, B-weighting was for levels between 55 and 85 dB, and C-weighting was for levels above 85 dB. These are shown in Fig. 1.5. Significantly, however, the A-weighting network is now used exclusively in most measurement standards and the mandatory noise limits.

Table 1.2 contains a listing of the corrections in decibels to be added algebraically to all frequency bands. The A-weighted sound pressure level is denoted as:

$$\text{``}L_{pA}, \text{ dB''} \quad \text{or} \quad \text{``}L_p, \text{ dBA''} \tag{1.35}$$

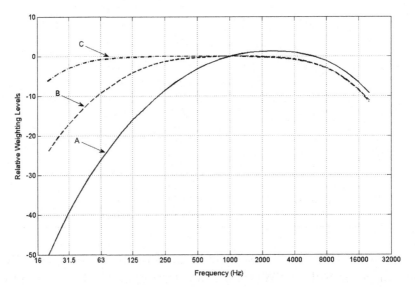

Fig. 1.5   Approximate electrical frequency response of the A-, B-, and C-weighted networks of sound level meters [12].

Table 1.2. Sound level conversion chart from flat response to A, B, and C weightings.

Frequency (Hz)	Octave band number	A weighting (dB)	B weighting (dB)	C weighting (dB)
20		−50.5	−24.2	−6.2
25		−44.7	−20.4	−4.4
31.5		−39.4	−17.1	−3.0
40		−34.6	−14.2	−2.0
50		−30.2	−11.6	−1.3
63	1	−26.2	−9.3	−0.8
80		−22.5	−7.4	−0.5
100		−19.1	−5.6	−0.3
125	2	−16.1	−4.2	−0.2
160		−13.4	−3.0	−0.1
200		−10.9	−2.0	0
250	3	−8.6	−1.3	0
315		−6.6	−0.8	0
400		−4.8	−0.5	0
500	4	−3.2	−0.3	0
630		−1.9	−0.1	0
800		−0.8	0	0
1000	5	0	0	0
1250		+0.6	0	0
1600		+1.0	0	−0.1
2000	6	+1.2	−0.1	−0.2
2500		+1.3	−0.2	−0.3
3150		+1.2	−0.4	−0.5
4000	7	+1.0	−0.7	−0.8
5000		+0.5	−1.2	−1.3
6300		−0.1	−1.9	−2.0
8000	8	−1.1	−2.9	−3.0
10,000		−2.5	−4.3	−4.4
12,500		−4.3	−6.1	−6.2
16,000		−6.6	−8.4	−8.5
20,000		−9.3	−11.1	−11.2

The former notation is more logical. However, the latter continues to be in wide use. Incidentally, symbol dBA is often written as dB(A).

The second column in Table 1.2 indicates octave numbers in popular use among professionals.

1.7 Logarithmic Addition, Subtraction and Averaging

The total sound pressure level of two or more incoherent sources of noise may be calculated as follows:

$$W_t = \sum_{i=1}^{n} W_i \tag{1.36}$$

or

$$L_{w,t} = 10 \log \left[\sum_{i=1}^{n} 10^{0.1 L_{w,i}} \right] \tag{1.37}$$

Here, n denotes the total number of incoherent sources like machines in a workshop or different sources of noise in an engine, etc. Similarly, the corresponding total SPL at a point is given by

$$L_{p,t} = 10 \log \left[\sum_{i=1}^{n} 10^{0.1 L_{p,i}} \right] \tag{1.38}$$

Incidentally, Eqs. (1.36)–(1.38) would also apply to logarithmic addition of SPL or SWL of different frequency bands in order to calculate the total level. The logarithmic addition of power levels or sound pressure levels has some interesting implications for noise control. It may easily be verified from Eqs. (1.37) and (1.38) that

$$100 \oplus 100 = 103 \, \text{dB}$$

$$100 \oplus 90 = 100.4 \, \text{dB}$$

$$x \oplus x = x + 3 \, \text{dB}$$

Similarly, 10 identical sources of x dB would add up to $x + 10$ dB. Perception wise,

3 dB increase in SPL is hardly noticeable.
5 dB increase in SPL is clearly noticeable; and
10 dB increase in SPL appears to be twice as loud.

Similarly, 10 dB decrease in SPL would appear to be half as loud, indicating 50% reduction in SPL. Therefore, it follows that:

(i) In a complex noisy situation, one must first identify all significant sources, rank them in descending order and plan out a strategy for reducing the noise of the largest source of noise first, and then only tackle other sources in a descending order.

(ii) While designing an industrial layout or the arrangement of machines and processes in a workshop, one must identify and locate the noisiest machines and processes together in one corner and isolate this area acoustically from the rest of the workshop or factory.

(iii) It is most cost-effective to reduce all significant sources of noise down to the same desired level.

Addition of the sound pressure levels of the incoherent sources of noise may be done easily by making use of Fig. 1.6 which is based on the following formula:

$$\Delta L \equiv L_{pt} - L_{p1} = 10 \log\{1 + 10^{-0.1(L_{p1} - L_{p2})}\}, \text{ dB} \qquad (1.39)$$

It may be noted that the addition ΔL to the higher of the two levels is only $0.4\,\text{dB}$ when the difference of the levels $(L_{p1} - L_{p2})$ equals $10\,\text{dB}$. Therefore, for all practical purposes, in any addition, if the difference between the two levels is more than $10\,\text{dB}$, the lower one may be ignored as relatively insignificant.

If one is adding more than two sources, one can still use Fig. 1.6, adding two at a time starting from the lowest, as shown in Fig. 1.6.

The concept of addition can also be extended to averaging of sound pressure level in a community location. Thus, the equivalent sound pressure level during a time period of 8 hours may be calculated as an average of the hourly readings; that is,

$$L_{p,8h} = 10 \log\left[\frac{1}{8}\sum_{i=1}^{8} 10^{0.1 L_{p,i}}\right], \text{ dB} \qquad (1.40)$$

This averaging is done automatically in an integrating sound level meter or dosimeter used in the factories in order to ensure that a worker is not subjected to more than 90 dBA of equivalent sound pressure level during an 8-hour shift. Similarly, one can measure L_d,

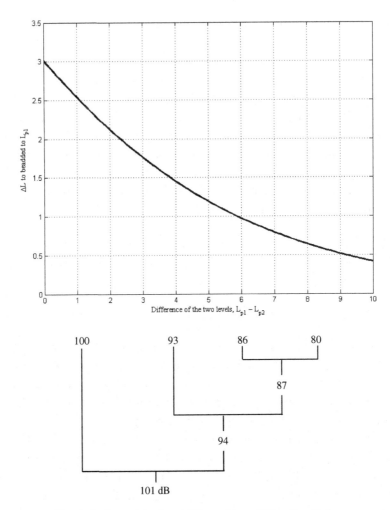

Fig. 1.6   Logarithmic addition of two SPLs, $L_{p1} \oplus L_{p2}$.

the daytime average (6 a.m. to 9 p.m.) and L_n, the night-time average (9 p.m. to 6 a.m.). Making use of the fact that one needs quieter environment at night, the day-night average (24-hour average) is calculated as follows:

$$L_{dn} = 10 \log \left[\frac{1}{24} \{ 15 \times 10^{0.1 L_d} + 9 \times 10^{0.1(L_n + 10)} \} \right], \text{ dB} \quad (1.41)$$

It may be noted that the L_n has been increased by 10 dBA in order to account for our increased sensitivity to noise at night.

Example 1.3. For a reasonably flat frequency spectrum with no sharp peaks or troughs, the band is nearly equal to the sum of the sound powers in the three contiguous one-third octave bands. Make use of this fact to evaluate sound pressure level of the 500 Hz band if the measured values of the 400, 500 and 630 Hz 1/3-octave bands are 80, 90 and 85 dB, respectively.

Solution. It may be noted from Table 1.1 that the frequency range of

400 Hz 1/3-octave band is 355–447 Hz,
500 Hz 1/3-octave band is 447–562 Hz,
630 Hz 1/3-octave band is 562–708 Hz, and
500 Hz 1/1-octave band is 355–710 Hz.

Thus, the 500-Hz octave band spans all three contiguous 1/3-octave bands. Therefore, making use of Eq. (1.37),

$$L_p(500\,\text{Hz octave band}) = 10\log(10^{80/10} + 10^{90/10} + 10^{85/10})$$
$$= 91.5\,\text{dB}$$

1.8 Directivity

Most practical sources of noise do not radiate noise equally in all directions. This directionality at distance r in the far field is measured in terms of a directivity index DI, or directivity factor DF, as follows:

$$DI_\theta(r) = L_{p,\theta}(r) - L_{p,av}(r) = 10\log(DF_\theta), \quad \text{dB} \qquad (1.42)$$

The average Sound pressure level $L_{p,av}$ is calculated from the total sound power level by

$$L_{p,av}(r) = L_W - 10\log(4\pi r^2) \qquad (1.43)$$

The average sound pressure level can be evaluated by averaging the measured sound pressure levels at different angles at the same distance (r) around the source, making use of the formula.

$$L_{p,av} = 10 \log \left[\frac{1}{n} \sum_{i=1}^{n} 10^{0.1 L_{p,i}} \right], \text{ dB} \qquad (1.44)$$

If all the levels around the machine are within 5 dB of each other, then instead of Eq. (1.44) one can take the arithmetic average of SPLs and add 1 dB to it in order to get a reasonably approximate value of the average sound pressure level.

Example 1.4. Sound pressure levels at four points around a machine are 85, 88, 92 and 86 dB when the machine is on. The ambient SPL at the four points (when the machine is off) is 82 dB. Calculate the average SPL of the machine alone (by itself).

Solution. Making use of Eq. (1.44),

$$L_{p,av}(\text{machine+ambient}) = 10 \log \left[\frac{1}{4} \left(10^{\frac{85}{10}} + 10^{\frac{88}{10}} + 10^{\frac{92}{10}} + 10^{\frac{86}{10}} \right) \right]$$

$$= 88.6 \text{ dB}$$

Incidentally, the arithmetic average of the SPLs at the four points works out to be,

$$\frac{1}{4}(85 + 88 + 92 + 86) = 87.7 \text{ dB}$$

So, the logarithmic average is quite close "to the arithmetic average plus 1 dB".

Now, logarithmically subtracting the ambient SPL we get,

$$L_{p,av}(\text{machine alone}) = 10 \log(10^{88.6/10} - 10^{82/10})$$

$$= 87.5 \text{ dB}$$

1.9 Loudness

Loudness index S is measured in terms of sones and the loudness level P in phons. Composite loudness index L and loudness level P

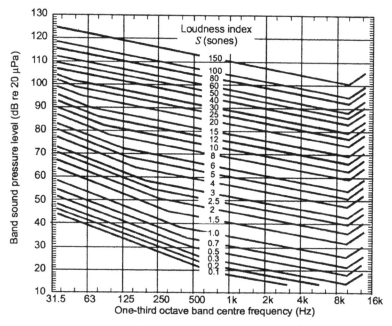

Fig. 1.7 Equal loudness index contours (adapted with permission from Ref. [3]).

are related to each other as follows:

$$S = 2^{(P-40)/10}, \quad P = 40 + 33.2 \log S \qquad (1.45)$$

The band loudness index for each of the octave bands is read from Fig. 1.7, and then the composite loudness index, L (sones) is determined by [4]:

$$L = S_{\max} + B \sum_i S_i \qquad (1.46)$$

where S_i is the loudness of the ith band and $S_{\max}$ is the maximum of these values. Constant $B = 0.3$ for octave band analysis and 0.15 for 1/3-octave band analysis. The summation does not include $S_{\max}$.

Example 1.5. If the measured values of the SPLs for the octave bands with mid-frequencies of 63, 125, 250, 500, 1000, 2000, 4000 and 8000 Hz are 100, 95, 90, 85, 80, 75, 70 and 65 dB, respectively, calculate the total loudness level in sones as well as phons.

Solution. The exercise is best done in a tabular form shown as follows.

Octave-band center frequency (Hz)	63	125	250	500	1000	2000	4000	8000
Octave-band level (dB)	100	95	90	85	80	75	70	65
Band loudness index (sones) (from Fig. 1.5)	28	25	22	19	17	14	13	11

The measured values of SPL are entered in the first row of the table. The band loudness index S_i for each octave band is read from Fig. 1.5 and entered in the second row of the table. It may be noted that the maximum value of the loudness index, $S_{\max}$, is 28 sones. Constant $B = 0.3$ for octave bands, Thus, making use of Eq. (1.46), the composite loudness index, L, is calculated as follows:

$$L = 28 + 0.3(25 + 22 + 19 + 17 + 14 + 13 + 11)$$

$$= 28 + 0.3 \times 121 = 64.3 \, \text{sones}$$

Now use of Eq. (1.45) yields the loudness level in phons:

$$P = 40 + 33.2 \log 64.3$$

$$= 100.0 \, \text{phons}$$

1.10 Noise Limits in India

The Ministry of Environment and Forests (MOEF) of the Government of India, on the advice of the National Committee for Noise Pollution Control (NCNPC) has been issuing Gazette Notifications prescribing noise limits as well as rules for regulation and control of noise pollution in the urban environment. These are summarized as follows.

1.10.1 *The noise pollution (regulation and control) rules, 2000*

These rules [5] make use of Table 1.3 for the ambient air quality standards. These are more or less the same as in Europe and USA.

(i) A loudspeaker or a public address system shall not be used except after obtaining written permission from the authority.

(ii) A loudspeaker or a public address system or any sound-producing instrument or a musical instrument or a sound amplifier shall not be used at night-time except in closed premises for communication within, like auditoria, conference rooms, community halls, banquet halls or during a public emergency.

(iii) The noise level at the boundary of the public place, where loudspeaker or public address system or any other noise source is being used, shall not exceed 10 dB(A) above the ambient noise standards for the area (see Table 1.3) or 75 dB(A) whichever is lower.

(iv) The peripheral noise level of a privately owned sound system or a sound-producing instrument shall not, at the boundary of the private place, exceed by more than 5 dB(A) the ambient noise standards specified for the area in which it is used.

(v) No horn shall be used in silence zones or during night-time in residential areas except during a public emergency.

Table 1.3. Ambient air quality standards in respect of noise [7].

Category of area/zone	Limits in *Leq* (dBA)	
	Day-time	Night-time
Industrial area	75	70
Commercial area	65	55
Residential area	55	45
Silence zone	50	40

Notes:
1. Day-time shall mean from 6:00 a.m. to 10:00 p.m.
2. Night-time shall mean from 10:00 p.m. to 6:00 a.m.

(vi) Sound-emitting firecrackers shall not be burst in silence zone or during night-time.

(vii) Sound-emitting construction equipment shall not be used or operated during night-time in residential areas and silence zones.

1.10.2 *Permissible noise exposure for industrial workers*

In keeping with the practice in most countries, India has adopted the international limit of 90 dBA during an 8-hour shift for industrial workers.

As shown in Table 1.4, for every 3 dB increase in the A-weighted sound level, the permissible maximum exposure has been reduced to half. Dosimeters have been provided to the factory inspectors and also to the traffic police. The technicians working on noisy machines or in noisy areas are provided with earmuffs or ear plugs, and are required to use them compulsorily.

The total daily dose, D, is given by

$$D = \frac{C_1}{T_1} + \frac{C_2}{T_2} + \cdots + \frac{C_i}{T_i} + \cdots + \frac{C_n}{T_n} \qquad (1.47)$$

where C_i is the total actual time of exposure at a specified noise level, and T_i is the total time of exposure permitted by the table above at that level.

Alternatively, if we want to evaluate the maximum time that a technician may be asked to work in a noisy environment without

Table 1.4.   Permissible noise exposure.

Duration/day (h)	Sound level (dBA) slow response
16	87
8	90
4	93
2	96
1	99
0.5	102
0.25 or less	105

risking him to over exposure, we may make use of an integrating sound level meter to evaluate the 8-hour average of the A-weighted SPL as follows:

$$L_{Aeq,8h} = 10 \log \left[\frac{1}{8} \int_0^T 10^{L_{pA}(t)/10} dt \right] \qquad (1.48)$$

where t is in hours. Then, the maximum allowed exposure time to an equivalent SPL, L_{Aeq}, $8h$, would be given by

$$T_{\text{allowed}} = 8/D \qquad (1.49)$$

where D, the daily noise dosage with reference to the base level criterion of 90 dBA is given by,

$$D = 2^{(L_{Aeq,8h}-90)/3} \qquad (1.50)$$

Here, constant 3 represents the decibel trading level which corresponds to a change in exposure by a factor of two for a constant exposure time ($10 \log 2 = 3$). For use in USA, this constant would be replaced with 5.

Example 1.6. The operator of a noisy grinder in an Indian factory is to be protected against over-exposure to the workstation noise by rotation of duties. The operator's ear level noise is 93 dBA near the grinder and 87 dBA in an alternative workplace. What is the maximum duration during an 8-hour shift that the worker may work on the grinder?

Solution. Referring to Table 1.4, T_1 for 93 dBA is 4 hours, and T_2 for 87 dBA is 16 hours.

Let the worker operate the grinder for x hours, and work in the quieter location for the remaining duration of $8 - x$ hours.

Use of Eq. (1.48) yields

$$1 = \frac{x}{4} + \frac{8-x}{16}$$

whence

$$x = \frac{8}{3} = 2.67 \, \text{hours}.$$

Therefore, the technician should not be made to operate the grinder for more than 2.67 hours during an 8-hour shift.

1.10.3 *Noise limit for diesel generator sets*

India has the problem of power scarcity although the government has set up a number of thermal power plants, hydropower plants and atomic power plants. Therefore, most of the manufacturers have their own captive power plants based on diesel engines. The relevant gazette notification of the MOEF prescribes as follows [6]:

- The maximum permissible sound pressure level for new diesel generator (DG) sets with rated capacity up to 1000 KVA shall be 75 dB(A) at 1 m from the enclosure surface, in free field conditions.
- The diesel generator sets should be provided with integral acoustic enclosure at the manufacturing stage itself.

Noise limits for diesel generator sets of higher capacity shall be as follows [7]:

- Noise from DG set shall be controlled by providing an acoustic enclosure or by treating the room acoustically, at the user's end.
- The acoustic enclosure or acoustic treatment of the room shall be designed for minimum 25 dB(A) insertion loss or for meeting the ambient noise standards, whichever is on the higher side. The measurement for Insertion Loss may be done at different points at 0.5 m from the acoustic enclosure/room, and then averaged.
- The DG set shall be provided with a proper exhaust muffler with insertion loss of minimum 25 dB(A).
- The manufacturer should offer to the user a standard acoustic enclosure of 25 dB(A) insertion loss, and also a suitable exhaust muffler with IL of at least 25 dB(A).

1.10.4 *Noise limit for portable gensets*

Most shops and commercial establishments have their kerosene-start, petrol-run portable gensets with power range of 0.5–2.5 KVA.

The relevant gazette notification for such small portable gensets prescribes as follows [7]:

- Sound power level may be determined by means of the Survey method [8] (see Fig. 1.4).
- The A-weighted sound power level of the source in the case of the direct method is calculated from the equation,

$$L_{WA} = L_{PA} - K + 10 \log(S/S_0) \tag{1.51}$$

where K is the environmental correction, $10 \log (1+4S/A)$, S is the area of the hypothetical measurement surface, m^2, $S_0 = 1$ m^2 and A is the room absorption, m^2 (see Eq. (6.13) in Chapter 6).

- The prescribed limit of sound power level of portable gensets is 86 dBA.
- This noise limit may necessitate use of acoustic hoods in most cases.

1.10.5 *Noise limit for firecrackers*

The Indian Society is a mix of different racial and religious communities. Each community has its festivals that are usually celebrated by means of sound emitting firecrackers. Most of the time, these crackers are fired in hand, and therefore the chance of body damage, particularly to the hearing for the players as well as on-lookers standing nearby, is high. Therefore, the government has mandated as follows [9]:

- The manufacture, sale or use of firecrackers generating noise level exceeding 125 dB(A) or 145 dB(C) peak at 4 m distance from the point of bursting shall be prohibited.
- For individual firecracker constituting a series (joined firecrackers), the above-mentioned limit is reduced by 5 log (N), where N = number of crackers joined together.
- The measurements shall be made on a hard concrete surface of minimum 5 m diameter or equivalent.
- The measurements shall be made in free field conditions, i.e., there shall not be any reflecting surface up to 15 m distance from the point of bursting.

1.10.6 *Noise limit for vehicles*

Environmental noise of vehicles is measured in a pass-by noise test [10] as shown in Fig. 1.8. The vehicle approaches line A–A at a steady speed corresponding to 3/4 times the maximum power speed of the engine. As the vehicle front end reaches point C_1, the accelerator is pushed to full open throttle position and kept so until the rear of the vehicle touches line B–B at point C_2. The maximum SPL reading is recorded at the two microphones locations shown in Fig. 1.8. The test is repeated with the vehicle moving in the opposite direction. This is repeated three times. The average of the peak SPL

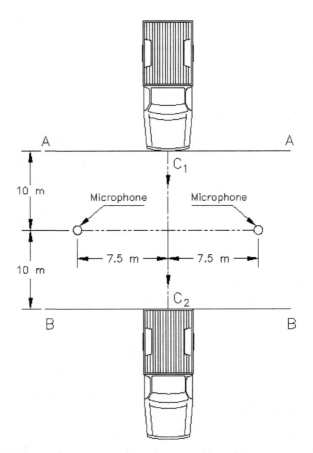

Fig. 1.8　Measurement of pass-by noise of an automobile [10].

Table 1.5. Noise limits for vehicles at manufacturing stage applicable since April 1, 2005 [10].

S. No.	Type of vehicle	Noise limits dB(A)
	Two-wheelers	
1	Displacement up to 80 cc	75
2	Displacement more than 80 cc but up to 175 cc	77
3	Displacement more than 175 cc	80
	Three-wheelers	
4	Displacement up to 175 cc	77
5	Displacement more than 175 cc	80
	Four-wheelers	
6	Vehicles used for the carriage of passengers and capable of having not more than nine seats, including the driver's seat	74
	Vehicles used for carriage of passengers having more than nine seats, including the driver's seat, and a maximum Gross Vehicle Weight (GVW) of more than 3.5 tonnes	
7	With an engine power less than 150 kW	78
8	With an engine power of 150 kW or above	80
	Vehicles used for carriage of passengers having more than nine seats, including the driver's seat\vehicles used for the carriage of goods	
9	With a maximum GVW not exceeding 2 tonnes	76
10	With a maximum GVW greater than 2 tonnes but not exceeding 3.5 tonnes	77
	Vehicles used for the transport of goods with a maximum GVW exceeding 3.5 tonnes	
11	With an engine power less than 75 kW	77
12	With an engine power of 75 kW or above but less than 150 kW	78
13	With an engine power of 150 kW or above	80

readings represents the pass-by noise of the vehicle. Detailed operating instructions as well as test conditions are given in Ref. [10].

Table 1.5 gives the pass-by noise limits for vehicles at the manufacturing stage [11] applicable since April 2005. These limits are similar to those prescribed in Europe since 1996. The limits are enforced by the Automotive Research Association of India (ARAI, Pune) during the type testing of the new vehicles for establishing their roadworthiness.

1.11 Masking

Often environmental noise masks a warning signal. Masking is the phenomenon of one sound interfering with the perception of another sound. This is why honking has to be considerably louder than the general traffic noise around. It has been observed that masking effect of a sound of a particular frequency is more at higher frequencies than at the lower frequencies. Thus, if we want to mask ambient sound of 500 Hz to 2000 Hz then we should introduce sound in the 500 Hz one-third octave band. In fact, the masking effect of a narrow band noise is more than that of a pure tone at the centre frequency of the band. The amount of masking is such that a tone which is a few decibels above the masking noise appears to be as loud as it would sound if the masking noise were not present. Masking can be put to effective use in giving acoustic privacy to intellectuals located in open cubicles in large call centers or similar large offices under the same roof. Playing of soft instrumental music in the background is enough to mask the conversation in the neighboring cubicles.

1.12 Sound Quality Parameters

The human auditory system is complex and reacts based on sound source characteristics. Generally, the human listening system tries to interpret the meaning of the sound after listening. Based on his/her knowledge base and experience, the cause for sound might be guessed. However, it might be misguided sometimes.

The human brain interprets the source of the cause and provides instructions to react. It acts like a processor. This study is a part of a Psychoacoustic study. The psychoacoustic study is a science to relate human perception with acoustics. Every working product can create a specific sound. This sound may be pleasant or unpleasant based on human perception.

Sound quality (SQ) study includes subjective (Jury) and objective tests. The purpose of a subjective test is to collect human perception of sound. Based on empirical equations, objective test data can be used to calculate SQ metrics like Loudness, Sharpness, Fluctuation of Strength, Roughness, Tonality, etc. These two tests have independent results. A proper analysis can establish the relation between these results [13].

The purpose of the subjective test is to capture human perception of different sounds. The subjective test consists of three major tasks. The first one is the preparation of sound samples for listening with proper instrumentation; the second one is jury evaluation methods, and the last one is the jury evaluation questionnaire.

The product sound samples can be recorded with actual product sounds in the field. It may not be possible to always conduct tests due to limited resources and complex interaction with other sounds. An alternative approach is to develop a procedure in the laboratory. This procedure may not match the test but provides guidelines or trend lines with proper assumptions. Binaural Recording sounds are closer to human listening. Recording with a single microphone loses the spatial information. Single-channel recording cannot relate to sound quality evolution. It is good for noise control applications. Test sample duration is an important aspect of subjective tests, and it is decided based on the type of Jury test.

The method of Jury evaluations must be selected so that the subjects' results represent customer opinion. Otto *et al.* [14] discuss jury evaluation methods for inexperienced subjects. The suggestion given by them is that no single method works best for every SQ analysis. Different subjective methods are Rank order, Response (rating) scale, Pair comparison method (evaluation task, detecting task, and similarity task), Semantic differential, and Magnitude estimation. Several methods have been developed, and a few popular methods are [13]:

1. Relevant Test (RT);
2. Semantic Differential Test (SDT);
3. Pair Comparison Test (PCT).

Psychoacoustics is the science that discusses the link between acoustic waves and human listening events. Traditional acoustic analysis pays attention to calculating sound pressure levels, and spectrum using weightings like A, B, C, and Linear. These calculations provide a rough estimation of perception. They do not provide aurally adequate information. So, there is a need to do psychoacoustic objective data analysis to calculate the psychoacoustic metrics and indices.

Major psychoacoustic metrics are Loudness, Specific Loudness, Roughness, Sharpness, and Fluctuations of Strength defined or calculated based on application. Each metric represents one type of sound behavior, but global representation is possible with psychoacoustic indices. For example, Relative sensory pleasantness is a function of Loudness, Roughness, Sharpness, and Tonality. It describes the global behavior of most sound-quality applications. Fastl and Zwicker [15] provide detailed information about psychoacoustic metrics and indices with formulas, etc.

1.12.1 *Pitch*

Pitch represents the subjective response to the frequency. The relationship between the frequency and pitch is nonlinear. The pitch depends on a pure tone's frequency and sound pressure level. Pure tones can be employed to ascertain the perception of pitch. The unit of pitch is "mel". One such approach involves quantifying the perception of "half-pitch". In this experimental scenario, participants are instructed to perceive a singular pure tone and modify the frequency of a subsequent tone to elicit a pitch precisely half that of the initial tone. The pitch follows linearly to the frequency at low frequencies and deviates at higher frequencies. The reference frequency is 125 Hz, and the pitch is 125 mel.

1.12.2 *Critical band rate*

The frequency and amplitude of any given sound affect the Human Auditory Response. Frequency plays a vital role in sound quality psychoacoustic metric calculations. The human ear combines sound stimuli situated closer to each other in the frequency domain of a single frequency band. Within this bandwidth, the loudness of tone and narrow band noise are perceived as the same. For a given frequency or a tone, a critical band is the smallest band of frequencies around which activates the basilar membrane [15]. The unit of critical bands is given as "bark". The critical band is simply a frequency resolution for the ear. The critical band approximates the frequency dependency to find objective psychoacoustic metrics. Fastl *et al.* [15] divided the audible frequency into 24 critical bands. Table 1.6 shows the critical bands with corresponding bandwidths.

Table 1.6. Critical frequency bands for the audible frequency range of the human ear [15].

Critical band rate z(Bark)	Frequency (Hz) (f_l, f_u)	Critical bandwidth Δf(Hz)	Critical band rate z(Bark)	Frequency (Hz) (f_l, f_u)	Critical bandwidth Δf (Hz)
0	0		12	1720	
		100			280
1	100		13	2000	
		100			320
2	200		14	2320	
		100			380
3	300		15	2700	
		100			450
4	400		16	3150	
		110			550
5	510		17	3700	
		120			700
6	630		18	4400	
		140			900
7	770		19	5300	
		150			1100
8	920		20	6400	
		160			1300
9	1080		21	7700	
		190			1800
10	1270		22	9500	
		210			2500
11	1480		23	12,000	
		240			3500
12	1720		24	15,500	

1.12.3 *Loudness*

Loudness is a hearing sensation that corresponds to sound level. It is defined as the level of a 1 kHz tone sound perceived as loud as the sound under consideration in a frontally incident plane field. Its value depends on sound intensity, duration, and temporal and spectral structure of sound.

The unit of loudness is "phon", and relative loudness is "sone". There are international standards to calculate the Loudness of steady-state sounds. The ISO 226-1987 standard is based on equal loudness contours for pure tones in free-field and narrow-band random noise in diffuse field. Loudness is calculated against the critical band for broadband and multi-tone sounds. It leads to the definition of Specific Loudness. ISO 532B standard is based on the Zwicker loudness assessment method, which can be considered for complex broadband noise, including pure tones. This method incorporates the masking effects. The details of loudness calculation are discussed in Section 1.9.

1.12.4 *Specific loudness*

Two sound signals with the same amplitude but different frequency contents would evoke different sensations of loudness in humans. The loudness perception of acoustic stimuli also depends on the spectral content. Specific Loudness is defined as the plot of the loudness curve against the critical bands. The unit of Specific Loudness is "sone/Bark".

Specific loudness is developed using the power law [15]. Specific Loudness (N') of a sound can be determined as

$$\frac{\Delta N'}{N'} = K\frac{\Delta E}{E} \tag{1.52}$$

where N' is a specific loudness, $\Delta N'$ is an increment in the specific loudness, E is excitation, ΔE is an increment in the excitation, and K is a proportionality constant.

By taking the threshold of quiet as a base or minimum level of excitation with a reference loudness, N_0', the final expression for Specific Loudness using a reference loudness N_0' [15], can be written as

$$N' = N_0' \left(\frac{E_{TQ}}{SE_0}\right)^K \left[\left(1 + \frac{SE}{E_{TQ}}\right)^K - 1\right] \qquad (1.53)$$

where E_{TQ} is an excitation level at the threshold of quiet, and S is the ratio between the audible test tone's intensity and the internal noise within the critical band at the test tone's frequency. E_0 is an excitation level corresponding to the reference intensity $I_0 = 10^{-12}$ W/m^2.

Fastl *et al.* [15] developed an approximation for the Specific Loudness in each critical band as follows:

$$N' = 0.08 \left(\frac{E_{TQ}}{E_0}\right)^{0.23} \left[\left(1 + 0.5\frac{E}{E_{TQ}}\right)^K - 1\right] \qquad (1.54)$$

$$N = \int_0^{24Bark} N' dz \qquad (1.55)$$

where N' is a function of the critical band (z).

1.12.5 *Sharpness*

Sharpness is a measure of the high-frequency content of a sound. Sharpness can be used partially to quantify sound quality. It is like a weighted loudness ratio. If any sound contains high-frequency content, that sound is perceived as sharper. Sharpness calculation is based on the specific loudness computation. The unit of sharpness is "acum". One acum is defined as a narrow band noise, one critical band wide at a center frequency of 1 kHz having a level of 60 dB. There are several methods to calculate sharpness [15].

According to Fastl *et al.* [15], sharpness is calculated as the weighted first moment of Specific Loudness (N'). The partial first moment is then weighted by the function $g'(z)$. The sum of these weighted partial moments is calculated and then divided by the Loudness (N).

$$Sharpness(s) = 0.11 \frac{\int_0^{24\text{Bark}} N'g'(z)dz}{\int_0^{24\text{Bark}} N'dz} \tag{1.56}$$

where $g'(z) = 1$ for $z < 14$,

$$g'(z) = 0.00012 \cdot z^4 - 0.0056 \cdot z^3 + 0.1 \cdot z^2 - 0.81 \cdot z + 3.51 \quad \text{for } z > 14.$$

1.12.6 *Fluctuation strength*

Sensation perceived due to very low-frequency modulation is termed fluctuation strength. Its unit is "vacil". One vacil is defined as a sine tone of 1 kHz with a level of 60 dB, amplitude modulated at a frequency of 4 Hz, and with a modulation depth of one. It occurs for modulation frequencies between 0 and 20 Hz. It represents the modulation of sound that is strongly audible. This fluctuation strength is developed based on temporal variation of the masking pattern. Studies show that the most significant sensation of fluctuation occurs at a modulation frequency of 4 Hz. Masking depth (ΔL) value varies inversely with increasing modulation. Fastl *et al.* [15] proposed an approximation for ΔL using maximum and minimum Specific Loudness in each critical band, as in Eq. (1.56).

$$\Delta L \approx 4 \log \left(\frac{N'_{\text{max}}}{N'_{\text{min}}} \right) \tag{1.57}$$

Fastl *et al.* [15] proposed fluctuation strength as given as below.

$$F \approx \frac{\Delta L}{((f_{\text{mod}}/4\,\text{Hz}) + (4\,\text{Hz}/f_{\text{mod}}))} \tag{1.58}$$

where ΔL is the depth of the temporal masking pattern, F is fluctuation strength, and f_{mod} is modulation frequency. Modulation frequency can be determined for any sound by examining its spectral content. Equation (1.58) shows that a 4 Hz modulation frequency is vital in describing fluctuation strength. Substituting Eq. (1.57) in Eq. (1.58) provides Eq. (1.59), which Fastl *et al.* [15] formed for empirical studies of fluctuation strength:

$$F = \frac{0.08 \int_0^{24\text{Bark}} \frac{4 \log\left(\frac{N'_{\max}}{N'_{\min}}\right) \text{dB}}{\text{dB/Bark}} dz}{(f_{\text{mod}}/4\,\text{Hz}) + (4\,\text{Hz}/f_{\text{mod}})} \tag{1.59}$$

1.12.7 *Roughness*

Roughness is an important parameter in the Sound Quality study. It quantifies rapid amplitude modulation of sound. Sensation of Roughness occurs in the frequency range (15–300) Hz. Roughness depends on the center frequency, modulation frequency, and modulation depth. When a sound changes its amplitude in the frequency range of 15–300 Hz, the human auditory system cannot recognize the total change due to temporal masking effects. Human auditory systems can only detect changes in excitation level or Specific Loudness at all places along the critical band rate scales.

At low modulation frequency (f_{mod}), Roughness is small, but masking depth (ΔL) is very high. At mid frequencies around 70 Hz, f_{mod} is high, ΔL is again small compared to low modulation frequencies. At high frequencies above 300 Hz, f_{mod} is very high, ΔL is significantly low due to the restriction of temporal resolution of the human auditory system. As ΔL value depends on the critical band rate, an approximation of Roughness R, which includes this dependence, is given as follows [15]:

$$R \approx f_{\text{mod}}\Delta L \tag{1.60}$$

Fastl *et al.* [15] gave a model for Roughness based on significant observation:

$$R = 0.3\frac{f_{\text{mod}}}{\text{kHz}} \int_0^{24\text{Bark}} \frac{20 \log\left(\frac{N'_{\max}}{N'_{\min}}\right) \text{dB}}{\text{dB/Bark}} dz \tag{1.61}$$

The unit of Roughness is asper. One asper is defined as a 1 kHz tone at 60 dB with 100% amplitude modulated at a modulated frequency of 70 Hz.

1.12.8 Tonality

The tonality of a sound describes the presence of tone in the sound. There are many tonality models available. Terhardt *et al.* [16] developed a method for calculating tonality, which includes frequency bandwidth and levels of all tonal components in a sound signal. The Aures model also describes the effect of bandwidth, frequency, and level of tonal component on the perception of tonality. This method first calculates the weighting function based on each tonal component's bandwidth, frequency, and total level, as given in the following equations [16]:

$$w_1(\Delta z_i) = \frac{0.13}{\Delta z + 0.13} \tag{1.62}$$

$$w_2(f_i) = \left(\frac{1}{\sqrt{1 + 0.2 \left(\frac{f_i}{700} + \frac{700}{f_i} \right)^2}} \right)^{0.29} \tag{1.63}$$

$$w_3(\Delta L_i) = \left(1 - e^{-\frac{\Delta L_i}{15}} \right)^{0.29} \tag{1.64}$$

where Δz_i is the bandwidth of the tonal component in bark, f_i is the frequency of the tone in Hz, ΔL_i is the excess level of the tonal component above the broadband masking noise. ΔL_i is given by Terhardt *et al.* [16] and Minard *et al.* [17] as

$$\Delta L_i = L_i - 10 \log_{10} \left\{ \left[\sum_{\substack{k=1 \\ k \neq i}}^{N} 10^{\frac{L_{EK(f_i)}}{20 \, \text{dB}}} \right]^2 + I_{NK}(f_i) + 10^{\frac{L_{TH(f_i)}}{10 \, \text{dB}}} \right\} \, \text{dB} \tag{1.65}$$

where L_i is the relative SPL of the ith spectrum sample, $L_{EK}(f_i)$ is the excitation level which is produced at the frequency f_i, $I_{NK}(f_i)$ is the noise intensity present in the critical band around the considered tonal components, $L_{TH}(f_i)$ is the threshold of hearing at the frequency (f_i). The previous weighting functions are combined to get an overall weighting function w_T.

$$w_T = \sqrt{\sum_{i=1}^{n}\left[w_1^{\frac{1}{0.29}(\Delta z_i)}w_2^{\frac{1}{0.29}(\Delta f_i)}w_3^{\frac{1}{0.29}(\Delta L_i)}\right]^2} \qquad (1.66)$$

Equation (1.66) represents a weighting factor due to the tonal characteristics of a sound. Another weighting factor, w_{Gr}, is based on the loudness signal-to-noise ratio.

$$w_{Gr} = 1 - \frac{N_{Gr}}{N} \qquad (1.67)$$

where N_{Gr} is the Loudness of the broadband noise component, and N is the total Loudness of the sound. Finally, Aures combined the total and noise weighting into single equation (1.68):

$$T = cw_T^{0.29}w_{Gr}^{0.79} \qquad (1.68)$$

Example 1.7. An electric vehicle motor produced distinct tones under three different experimental conditions:

(i) Tones at 1000 Hz and 1004 Hz,
(ii) Tones at 1000 Hz and 1070 Hz, and
(iii) Tones at 1000 Hz and 1200 Hz.

Comment on the human auditory system's perception of these three conditions.

Solution. Referring to Table 1.6, critical bands, and bandwidths for frequencies between 1000 Hz and 1200 Hz are illustrated in Fig. 1.9.

For Case (i), where the difference between the tones is 4 Hz (within the critical bandwidth), humans perceived a single tone with slight low-frequency modulation or beating. This perception aligns with what's known as Fluctuation of Strength.

In Case (ii), the difference between the tones was 70 Hz (still within the critical bandwidth). Despite this, humans perceived it as a single tone with high-frequency modulation, a phenomenon termed Roughness.

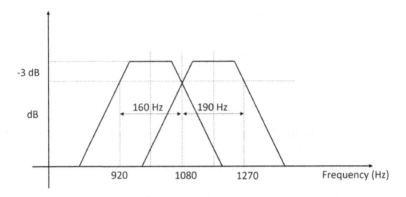

Fig. 1.9   Critical band rate and increase in bandwidth for high frequency.

Case (iii), where the difference between the tones was 200 Hz (beyond the critical band bandwidth), the human ear perceived it as two distinct tones.

References

[1] Munjal, M. L., *Acoustics of Ducts and Mufflers*, Second Edition, Wiley, Chichester, UK (2014).

[2] Irwin, J. D. and Graf, E. R., *Industrial Noise and Vibration Control*, Prentice Hall, Englewood Cliffs (1979).

[3] Bies, D. A. and Hansen, C. H., *Engineering Noise Control*, Fourth Edition, Spon Press, London (2009).

[4] *Procedure for the Computation of Loudness of Noise, American National Standard USA S3.4-1968*, American National Standards Institute, New York (1968).

[5] The Noise Pollution (Regulation and Control) (Amendment) Rules, 2010, MOEF Notification S.O. 50(E), The Gazette of India Extraordinary (11 January 2010).

[6] MOEF Notification G.S.R. 371(E): Environment (Protection) Second Amendment Rules (2002).

[7] MOEF Notification G.S.R. 742(E): Environment (Protection) Amendment Rules (2000).

[8] Acoustics — Determination of sound power levels of noise sources using sound pressure — Survey method using an enveloping measurement surface over a reflecting plane, ISO 3746: 1995(E), International Standards Organization, (1995).

[9] MOEF Notification G.S.R. 682(E): Environment (Protection) Second Amendment Rules (1999).

[10] Measurement of noise emitted by moving road vehicles, Bureau of Indian Standards, IS: 3028-1998, New Delhi (1998).

[11] MOEF Notification G.S.R. 849(E): Environment (Protection) Second Amendment Rules (2002).

[12] Joint Departments of the Army, Air Force and Navy, TM 5-805-4/AFJMAN 32-1090, Noise and Vibration Control (1995).

[13] Mahanta, T. K., Sound Quality Evaluation of Automotive Horn, Ph.D. Thesis, Indian Institute of Technology Hyderabad (2018).

[14] Otto, N., Amman, S., Eaton, C., and Lake, S., Guidelines for jury evaluations of automotive sounds. *Sound and Vibration*, 35(4), (2001) 24–27.

[15] Fastl, H. and Zwicker, E., *Psychoacoustics: Facts and Models*, Vol. 2, Springer Science and Business Media (2007).

[16] Terhardt, E., Stoll, G. and Seewann, M., Algorithm for extraction of pitch and pitch salience from complex tonal signals. *The Journal of the Acoustical Society of America*, 71(3), (1982) 679–688.

[17] Minard, A. and Boussard, P., Signal-based indicators for predicting the effect of audible tones in the aircraft sound at takeoff. *INTER-NOISE and NOISE-CON Congress and Conference Proceedings, INCE*, 253(2), (2016) 6410–6419.

Problems

1.1. Often, an approximate value of characteristic impedance ($Z_0 = \rho_0 c$) of air is taken as a round figure of $400 \, \text{kg}/(\text{m}^2\text{s})$. What temperature does this value correspond to? Adopt the mean sea level pressure.

[Ans.: $40°$C]

1.2. What are the amplitudes of the particle velocity and sound intensity associated with a plane progressive wave of $100 \, \text{dB}$ sound pressure level? Assume that the medium is air at mean sea level and $25°$C.

[Ans.: $6.9 \, \text{mm/s}$ and $9.76 \, \text{mW/m}^2$]

1.3. A bubble-like pulsating sphere is radiating sound power of 0.1 Watt at 500 Hz. Calculate the following at a far field point located $1.0 \, \text{m}$ away from the centre of the sphere.

(a) intensity level and sound pressure level

(b) rms value of acoustic pressure

(c) rms value of the radial particle velocity

(d) phase difference between pressure and velocity

[**Ans.: (a) both 99.0 dB, (b) 1.78 Pa, (c) 4.37 mm/s,**
(d) 6.3°]

1.4. Sound pressure levels at a point near a noisy blower are 115, 110, 105, 100, 95, 90, 85 and 80 dB in the 63, 125, 250, 500, 1000, 2000, 4000 and 8000 Hz octave bands, respectively. Evaluate the total L_p and L_{pA}.

[**Ans.: $L_p = 116.6$ dB, $L_{pA} = 102.4$ dBA**]

1.5. Sound pressure levels measured in free space at 12 equispaced locations on a hypothetical hemi-spherical surface of radius 2m around a small portable genset lying on an acoustically hard floor are 70, 72, 74, 76, 78, 80, 81, 79, 77, 75, 73 and 71 dB, respectively.

(a) Evaluate the sound power level of the portable genset.

(b) What is the directivity factor for the locations where SPL is maximum and where SPL is minimum?

[**Ans.: (a) 90.8 dB, (b) 5.25 and 0.417**]

1.6. At a point in a particular environment, the octave-band sound pressure levels are 80, 85, 90, 85, 80, 75, 70 and 65 dB in the octave bands centered at 63, 125, 250, 500, 1000, 2000, 4000 and 8000 Hz, respectively. Evaluate the loudness index in sones for each of these eight octave bands, and thence calculate the total loudness level in sones and phons.

[**Ans.: 49.6 sones and 96.5 phons**]

1.7. If the hourly A-weighted ambient SPL in an Indian workshop during an 8-hour shift are recorded as 85, 86, 87, 88, 90, 93, 93, 91 dBA, evaluate (a) the daily noise dose of the technicians employed in the workshop, and (b) the maximum permissible exposure time according to the industrial safety standards.

[**Ans.: (a) 1.012 and (b) 7.9 hours**]

Chapter 2

Acoustic Measurements

2.1 Introduction

The purpose of acoustic measurement is to quantify and analyze sound in various environments. It can include measuring the amplitude, frequency, duration of sounds and identifying and analyzing specific types of noise. The typical acoustic and vibration measurement chain is shown in Fig. 2.1.

The acoustic and vibration signals are sensed and transformed into electrical energy by means of microphones and accelerometers. The sensor output electrical signals are conditioned for appropriate gain and impedance matching for proper signal transfers with a high signal-to-noise ratio through the cables. The selection of cables and connectors depends on the transducers and data acquisition (DAQ) system. The processed digital data is displayed and controlled according to the application.

Product testing involves measuring and analyzing the sound emitted by machines such as appliances, vehicles, and industrial equipment to ensure they meet the design and safety standards. Acoustic measurement equipment can have inherent uncertainty, which can affect the accuracy of measurements. Regular calibration is necessary to ensure the equipment functions correctly and produces accurate results.

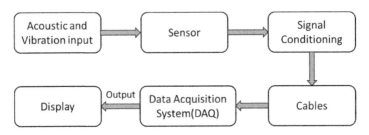

Fig. 2.1   Acoustic and vibration measurement chain.

Acoustic measurement is used in a variety of industries and applications as follows:

Industrial noise control: This involves measuring and analyzing the level and quality of noise in industrial environments to identify noise sources and implement noise control measures. Acoustic measurement can be challenging due to complex sound sources' presence in an industrial environment. The sound from multiple sources can be hard to separate and accurately measure. Some areas may be difficult to access for measurement, such as inside industrial machinery or in large, open spaces.

Building acoustics: This involves measuring and analyzing the sound transmission properties of buildings and other structures to ensure they meet design and safety standards and assessing and improving the acoustic comfort of a room or space for specific purposes, such as speech intelligibility, speech privacy, and sound insulation.

Environmental noise monitoring: This involves measuring and analyzing the level and quality of noise in each area to determine its impact on the environment and human health. In specific industrial sectors, compliance with regulations and standards, such as Occupational Safety and Health Act (OSHA), National Institute of Occupational Safety and Health (NIOSH), and International Standard Organization (ISO). Background noise can interfere with the accuracy of measurements and make it difficult to distinguish between different sound sources.

Music and audio production: This involves measuring and analyzing the sound and noise in music and audio production to ensure optimal sound quality. How humans perceive music can vary, making it difficult to establish a standard for measuring audio signals.

Acoustic measurement can be conducted using various techniques and instruments, such as sound level meters (SLMs), microphone arrays, and octave and 1/3 octave band analyzers. The choice of method and instrument depends on the specific application and the type of data that needs to be collected and analyzed.

2.2 Microphones

Acoustic pressure is measured using microphones. It converts acoustic pressure into electrical output. Several types of microphones are available for measurements, each with unique characteristics and applications.

The most commonly used microphones are:

(1) Condenser microphone;
(2) Electret microphone;
(3) Piezoelectric microphone;
(4) Dynamic microphone.

2.2.1 *Condenser microphone*

A condenser microphone is a type of microphone that uses a capacitor principle to convert sound waves into an electrical signal. The capacitor in a condenser microphone comprises two metal plates; one is fixed as a backing plate, and the other is a diaphragm that vibrates when sound waves strike it as shown in Fig. 2.2. The movement of the diaphragm changes the distance between the two plates (h_0), which in turn changes the capacitance of the capacitor. This change in capacitance is then converted into an electrical signal by an attached preamplifier, which can then be amplified and recorded. Condenser microphones are generally considered more sensitive, relatively flat frequency response, and accurate than dynamic microphones but

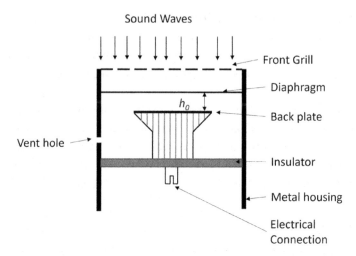

Fig. 2.2   Schematic diagram of a condenser microphone.

require external power. These microphones are sensitive to dust and moisture.

The bias voltage, or pre-polarizing voltage, is a constant voltage applied to a device, such as a microphone or a transistor, to establish a steady-state operating point. In the case of a condenser microphone, the bias voltage is applied to the backplate of the microphone to create a steady charge on the diaphragm. The bias voltage is normally supplied by a power source, such as a battery or a phantom power supply connected to the microphone.

The fixed charge on the backplate causes voltage variation across the plates inversely proportional to the capacitance as given in Eq. (2.1)

$$V = \frac{Q}{C} \tag{2.1}$$

where V is voltage (variation of voltage in microphone signal), Q is back plate charge (constant), and C is capacitance that changes as the diaphragm moves.

The bias voltage is necessary for properly operating a condenser microphone because the diaphragm must have a steady charge to respond to changes in sound pressure and create an electrical signal that can be amplified and processed. The microphone won't work properly without bias voltage, and the resulting audio signal would be distorted or weak.

It is important to note that different microphones have different bias voltage requirements. The typical bias voltage is 200 V. It is essential to check the microphone's specifications and ensure that the bias voltage is within the acceptable range.

2.2.2 *Electret microphone*

An electret microphone is a type of condenser microphone that uses a permanent electrostatic charge on a special material called an electret to establish a steady bias voltage on the diaphragm. It eliminates the need for an external power source to supply bias voltage. Hence, these microphones are called "self-polarized" microphones.

In an electret microphone, a thin layer of electret material is placed on or near the diaphragm. This material has a permanent electrostatic charge, creating a steady bias voltage on the diaphragm. When sound waves hit the diaphragm, it vibrates and changes its charge, creating a small electrical signal that is amplified and sent to the DAQ system. These microphones have higher capacitance as compared to condenser microphones. Figure 2.3 shows the schematic diagram of the electret microphone.

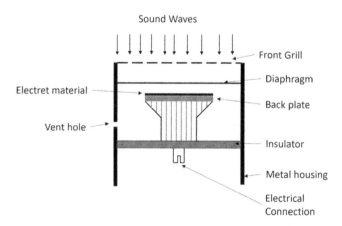

Fig. 2.3   Schematic diagram of an electret microphone.

Electret microphones are popular due to their low cost, small size, and high performance. They are common in consumer electronic devices such as mobile phones, laptops, and digital voice recorders. Depending on the design, they can be either omnidirectional or directional and are known for their high sensitivity and good frequency response.

It's important to note that electret microphones are not as sensitive as a traditional condenser microphone, but they are still widely used due to their low cost and ease of use.

2.2.3 *Piezoelectric microphone*

Piezoelectric microphones work by using the piezoelectric effect. This effect occurs when certain materials, such as lead-titanate, lead-zirconate, or barium-titanate, produce an electric charge when subjected to mechanical stress. In a piezoelectric microphone, the piezoelectric element is put before a sound pressure wave, which applies mechanical stress to the element as shown in Fig. 2.4. This stress generates an electrical signal proportional to the sound pressure wave, which is then processed and amplified to produce an output signal. The output signal can be used for recording or

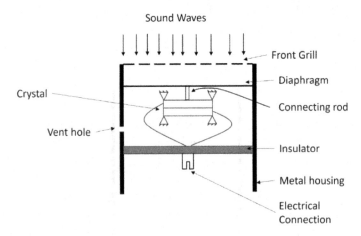

Fig. 2.4   Schematic diagram of piezoelectric microphone.

transmission purposes. These microphones are sensitive to vibration. Piezoelectric microphones are also called ceramic microphones.

2.2.4 *Dynamic microphone*

Dynamic microphone's working principle is that a moving coil of wire generates an electrical signal in response to sound waves. A flexible diaphragm is positioned before a magnetic assembly and vibrates in response to incoming sound waves as shown in Fig. 2.5. The voice coil is a wire coil attached to the diaphragm and suspended within a magnetic field. The magnet provides a magnetic field interacting with the moving voice coil to generate an electrical signal. This electrical signal is then transmitted to external circuitry for amplification and processing.

Dynamic microphones are robust and can handle high sound pressure levels (SPLs) without distorting, making them well-suited for live performances and other loud environments. These have low output impedance and can be used with long cables. They also tend to be relatively inexpensive compared to other types of microphones.

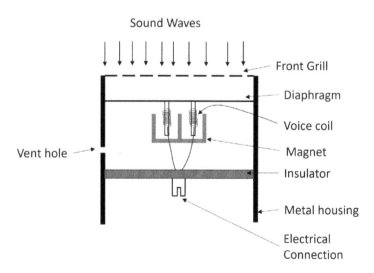

Fig. 2.5   Schematic diagram of dynamic microphone.

2.3 Microphone Specifications

There are several key specifications to consider when looking at a microphone. These are as follows:

Sensitivity: This measures the microphone's ability to convert sound energy; the higher the sensitivity, the more electrical output from the microphone for same input of sound. Microphone sensitivity is essentially the ratio of electrical output to acoustical input. In logarithmic units, it is defined as

$$S_m = 20 \log \left(\frac{E}{E_{ref}} \frac{p_{ref}}{p} \right), \ \text{dB} \qquad (2.2)$$

E_{ref}, the reference voltage is generally 1 volt and p_{ref} the reference pressure which can be 1 Pa or 0.1 Pa (1 microbar). Equation (2.2) may be rearranged as

$$S_m = 20 \log(E) - L_p + 94, \ \text{dB} \qquad (2.3)$$

where E is in volts and L_p is the SPL (re $20\,\mu$Pa) on the microphone. Typical values of microphone sensitivities range between -25 and 60 dB re 1V/Pa, and linear units for microphone sensitivity is mV/Pa, as shown in Fig. 2.6.

The sensitivity is a function of the size of the microphone, and it increases as the diaphragm size increases. The standard sizes of microphones are $1''$, $1/2''$, and $1/4''$. The microphone's sensitivity is measured using calibration with a known source, or comparison with a reference microphone. The chosen environment for the comparison method is an anechoic chamber, reverberation room, or plane wave tube. In a comparison method, initially, measure the electrical output response (V_0) with a standard microphone having sensitivity (S_0) for a given source and environment, and then replace the standard microphone with the calibrated microphone and the measure the output response (V_m). The microphone sensitivity (S_m) is then calculated from Eq. (2.4).

$$S_m = \frac{V_m}{V_0} S_0 \qquad (2.4)$$

Pistonphone calibrators are used to measure the absolute sensitivity of the microphone. The measurements are performed at a tonal frequency, part of the flat frequency response range. Free-field corrections are applied for frequencies outside the flat frequency range.

Frequency response: This measures the microphone's ability to replicate different frequencies accurately. It is typically measured in Hertz (Hz) and is often displayed by a graph showing the microphone's response to different frequencies. Some microphones have a flat frequency response, meaning they accurately capture all frequencies within their range. Others have a boosted or cut response, which can affect the quality of the captured sound. The presence of a microphone in the measurement field should not disturb the field. However, if the diameter of the microphone (d) is in the order of wavelength (λ), then diffraction effects are significant ($\sim d > \lambda/20$). So, the microphone size limits the higher frequency of measurement.

Free-field correction: The difference in pressure response between the pressure on the surface (P_r) of the microphone in a free field and pressure that would exist at the microphone location without the microphone in place (P_0) is defined as the free-field correction. It is a function of frequency and angle of incidence. The free-field correction values are higher at high frequencies and for microphones with a protective grid. The frequency response of the microphone deviates from the flat response due to the free-field corrections. Figure 2.6 shows the sensitivity change as a function of frequency for various microphone sizes.

Directionality: This refers to the microphone's pickup pattern, which can be omnidirectional, unidirectional, or bidirectional. Omnidirectional microphones pick up sound from all directions, unidirectional microphones pick up sound from one direction, and bidirectional microphones pick up sound from two directions.

Impedance: This measures the microphone's resistance to electrical current and is typically measured in ohms. Low-impedance microphones are easier to drive and less affected by long cables, while high-impedance microphones require more power.

Dynamic range: Dynamic range refers to a range of SPLs that a microphone can handle without distorting the output signal.

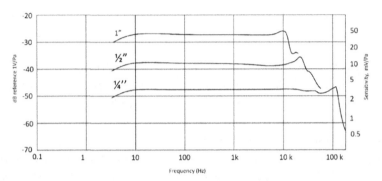

Fig. 2.6 Sensitivity variation for different microphone sizes as a function of frequency.

Microphones with a high dynamic range can be better to handle loud sounds without distorting them.

Noise floor: This refers to the amount of background noise generated by the microphone. It is measured in decibels (dB); the lower the noise level floor, the better the microphone is at isolating the desired sound.

Power requirement: Some microphones need an external power source to operate. It is called "phantom power" or "plug-in power." Microphones with low power consumption are ideal for use in battery-powered devices.

Connector: This refers to the type of connector the microphone uses to connect to an audio recording or processing device. Common connector types include BNC, LEMO, Micro-dot, XLR, TRS, and USB.

Size and weight: This is the physical dimension and weight of the microphone. It might be important if you use the microphone in a portable setup. Different sizes of microphones are available, each with unique characteristics and applications.

It's important to note that different types of microphones have different strengths and weaknesses, and the best microphone for a given application depends on the specific requirements of the situation. The relative performance of these microphones is given in Table 2.1.

The important specifications of the PCB microphone (Model: 378B02) used are tabulated in Table 2.2 and the half-inch condenser microphone is shown in Fig. 2.7.

Table 2.1.   Basic microphone types and characteristics.

	External-polarized condenser microphone	Self-polarized condenser (electret) microphone	Piezoelectric (ceramic) microphone	Dynamic microphone
Sensing element	Capacitor	Capacitor	Piezoelectric	Magnetic coil
Typical frequency range	2 Hz–20 kHz	2 Hz–20 kHz	20 Hz–10 kHz	25 Hz–15 kHz
Cables & connectors	7-wire connection, external 200 V supply required, LEMO	IEPE 2-wire connection, BNC connector	IEPE 2-wire connection, BNC connector	Balanced cable, XLR Connector
Temperature and humidity stability	Good	Fair	Excellent	Good
Dynamic range	Excellent	Excellent	Good	Good
Cost	High	Low	Low	Low

Table 2.2.   Specifications of microphone (reference: PCB Piezo electronics).

S. No.	1/2″ pre-polarized condenser microphone	Magnitude
1	Sensitivity	48.7266 mV/Pa ($\pm$ 1.5 dB)
2	Frequency range	3.75 Hz–20 kHz ($\pm$ 2dB)
3	Dynamic range	137 dB re 20 μPa
4	Constant current excitation	2–20 mA
5	Noise floor	15.5 dB(A) re 20 μPa
6	Connector	BNC Jack

Fig. 2.7   Half-inch (12.7 mm) condenser microphone (courtesy Hottinger Brüel & Kjaer reference).

2.4 Preamplifier

A preamplifier (preamp or pre) is an electronic device that amplifies a weak electrical signal before sending it to a power amplifier. The microphone or other source signal is typically too weak to drive a power amplifier directly. Hence, the preamp boosts the signal to a level that the power amplifier can handle. Preamplifier has high input impedance and small parallel capacitance. It allows the signal amplification to be much higher before sending it to a speaker or other output device. The preamplifiers have a low impedance output for driving long cables to a portable noise analyzer or DAQ systems. Preamps can also adjust the gain, tone, and other signal characteristics before it is amplified. Condenser microphones have low capacitance, and the preamplifier should be close to the transducer to avoid voltage loss over the cable. Piezoelectric and electret microphones have higher capacitance compared to condenser microphones. Hence, these microphones can connect to longer cables between the sensor and preamplifier.

2.5 Data Acquisition System

DAQ system hardware acts as the interface between a computer and signals from the sensors. DAQ involves gathering signals from measurement sources and digitizing the signals for storage, analysis, and presentation on a PC. It is a device that digitizes incoming analog signals so that a computer can interpret them. The three critical components of a DAQ device used for measuring a signal are the signal conditioning circuit, an analog-to-digital converter (ADC), and a computer bus.

The essential components of a DAQ system include:

Sensors: These devices are used to measure various physical or electrical parameters. They can include thermocouples, load cells, accelerometers, microphones, and other types of sensors.

Signal conditioning: This is the process of preparing the signals from the sensors for further processing. It can include amplifying, filtering, or converting the signals to a different format.

DAQ hardware: This device converts the conditioned sensor signals into digital data that a computer can process.

DAQ software: This software runs on the computer, controls the DAQ hardware, and processes the data.

Output devices: These devices display or store the data which the DAQ system collects.

2.6 Connectors

Depending on the specific application and requirements several types of connectors can be used in the measurement systems. Some common types of connectors used in the acoustic and vibration measurement are shown in Fig. 2.8. BNC is a type of connector that is commonly used in laboratory and scientific applications. It provides a secure and reliable connection for low-frequency signals. An acronym BNC is formed from the initial letters of the pattern, which is **B**ayonet, and the names of the inventors **N**eill-**C**oncelman. The TNC connector is a threaded version of the BNC connector. LEMO is the name of an electronic and fibre optic connector manufacturer based in Switzerland. Microdot is a co-axial connector which is used for accelerometers and microphones.

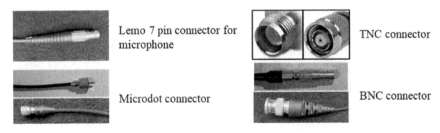

Fig. 2.8   Various connectors used in acoustic and vibration measurements.

2.7 Sound Level Meter (SLM)

A SLM is a portable device to measure a specific environment's SPL. It typically consists of a microphone, pre-amplifier, weighting networks, an amplifier, rectifier, a detector, and a display as shown in Fig. 2.9.

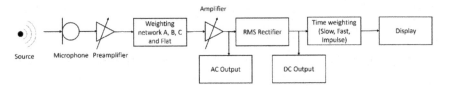

Fig. 2.9   Block diagram of SLM.

The SLM's microphone converts the sound pressure waves into an electrical signal. The electrical signal produced by the microphone is at a very low level, so a preamplifier amplifies it before sending the ADC. Signal processing includes applying frequency and time weightings to the signal as specified by international standards such as IEC 61672-1, to which SLMs conform. The display shows the sound level in decibels, typically with a descriptor showing the selected combination of time and frequency-weighting. The signal may also be available at the output sockets, in either AC or DC form, for connection to external instruments, such as a DAQ system, to provide a record and further processing. Some SLMs also have built-in memory and data logging capability, allowing users to record measurements over time.

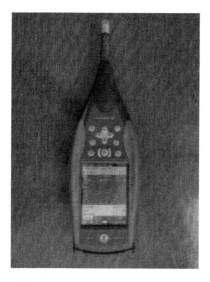

Fig. 2.10   Class 1 SLM with microphone (courtesy Hottinger Brüel & Kjaer).

Figure 2.10 shows the handheld analyzer with the microphone. The SLM should be able to capture frequency and the time-varying nature of the amplitude.

Several parameters can be measured by an SLM, including:

SPL: The amplitude of the sound is the most measured parameter and is typically measured in dB.

Time weighting: Time weighting refers to how the SLM measures and responds to sound over time. Sound pressure measurement has three main types of time weighting which are slow, fast, and impulse.

1. **Slow (S) time weighting:** This type of time weighting measures the sound level over a period; typically, its value is 1 s. It is useful for measuring steady-state noise, such as the noise from a running engine or a fan.
2. **Fast (F) time weighting:** This time weighting measures the sound level over a shorter period. Typically, its value is 125 ms. It helps measure transient noise, such as the noise from a hammer striking a nail.

3. **Impulse (I) time weighting:** This time weighting measures the sound level over an extremely short period. Typically, its value is 35 ms. It helps measure short-duration, high-level impulse noise like gunfire or explosions.

Most SLMs allow the user to select the time weighting, and the choice of time weighting would depend on the specific application and the type of noise being measured. It is also important to note that different standards may have different requirements for time weighting, so it is essential to refer to the appropriate standard for the specific measurement application.

Frequency weighting: Frequency weighting is a method used in sound level measurement to adjust the sensitivity of the measurement to match the human ear's sensitivity to different frequencies.

The most common frequency weighting is A-weighting, but there are also B, C, and Z weighting.

A-weighting: This weighting is the most widely used, approximating the human ear's sensitivity to different frequencies. The human ear is less sensitive to low and high frequencies than mid-frequencies. It is used for measuring general environmental noise and for measuring noise in the workplace. A-weighted network curve is established for below 55 dB SPL spectrum.

B-weighting: This weighting is a modification of the A-weighting and is intended for measurements of low-frequency noise, such as that produced by wind turbines, vehicles, or transportation. The B-weighted network curve is established for SPLs between 55 and 85 dB.

C-weighting: This weighting measures peak SPLs, such as those produced by impact or impulse noise. The C-weighted network curve is established for SPLs above 85 dB.

Z-weighting: This weighting measures impulse noise, such as gunfire or explosions. It has a flat frequency response and is not adjusted to the human ear's sensitivity. IEC 61672 defines the Z weighting.

A, B, and C frequency weighting was intended to mimic the human ear response to low, medium, and high noise levels, respectively. However, the current practice uses A and C weighting for all levels.

2.7.1 *Sound level meter specifications*

The specifications of a SLM typically include the following:

Class of SLMs: IEC61672/1-2013 standard classifies the SLM into two classes according to the accuracy.

1. **Class 1 SLMs:** These are highly accurate and precise instruments used in laboratory settings and for compliance measurements. They are typically more expensive and complex than other types of SLMs. The accuracy of class 1 SLM is ± 0.5 dB, and repeatability is ± 0.1 dB in an indoor environment.
2. **Class 2 SLMs:** These are less accurate and less precise than Class 1 SLMs but still meet the requirements of most standards for general-purpose measurements. The accuracy of class 2 SLM is ± 1 dB in an indoor environment. However, the measured precision in outdoor measurement may be more than 1 dB.

Microphone: The type of microphone used in the SLM depends on accuracy class and environment, such as a pressure or a free-field microphone. The sensitivity of the microphone is characterized according to the environment and angle of sound incidence. SLM microphones are also calibrated for random incidence because of the unknown sound incidence angle. It is suggested to check the microphone calibration chart provided by the manufacturer. The microphone orientation is to be chosen according to the manufacturer's specifications. Figure 2.11 shows the various microphone orientations with respect to source in free-field and diffuse field. The random incidence microphone is oriented at an angle of 70° to 90° (grazing incidence) from the noise source. The microphone axis is pointed at the noise source for a perpendicular direction.

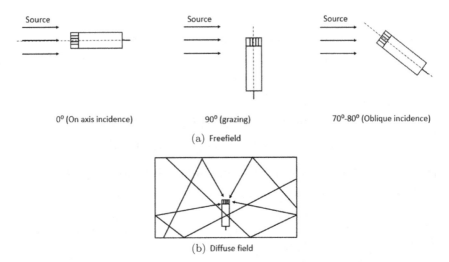

(a) Freefield

(b) Diffuse field

Fig. 2.11 Microphone orientation with respect to sound source: (a) Free field and (b) diffuse field.

Frequency range: The range of frequencies the SLM can measure, typically measured in Hz.

Dynamic range: The difference between the highest and lowest measurable sound levels, typically dB.

Accuracy: The degree of precision and correctness of the measurements, typically measured in dB.

Time weighting: Time weighting refers to how the SLM measures and responds to sound over time. It can be Slow, Fast, or Impulse.

Frequency weighting: The frequency weighting adjusts the sensitivity of the measurement to match the human ear's sensitivity to different frequencies. It can be A, B, C or Z.

Display: The type of display and the information it provides, such as the SPL, L_{eq}, L_{max}, L_{min}, and L_{peak}.

Data logging: The capability of the SLM to store measurement data and the way of data transfer, such as USB, Bluetooth, or Wi-Fi.

Battery life: The duration of the battery life.

Properly planning to do measurements with a SLM is essential for better accuracy and repeatability. A few points must be considered before measurement, although the list is not all-inclusive.

- Purpose of the measurements;
- Environment (indoor, outdoor, diffusive, free-field, and pressure-field, etc.), background noise, wind speed, atmospheric pressure, temperature;
- Measurement duration;
- Type of sound (steady continuous, steady intermittent, fluctuating continuous, fluctuating intermittent, impulse continuous, impulse intermittent);
- Source characteristics (directivity, frequency, amplitude, etc.), operating conditions;
- Type of instruments and their calibration process.

Based on the temporal characteristics, the type of sound is categorized as continuous and intermittent. The typical sound type and possible settings in SLM are mentioned in Table 2.3.

2.8 Dosimeter

A dosimeter is a device used to measure and record an individual's exposure to noise or other forms of sound over a period. The individual wears it and typically measures the SPL and the duration of the exposure. The measurement results are usually stored in the dosimeter's memory and can be downloaded and analyzed later.

A dosimeter typically consists of a microphone, pre-amplifier, an amplifier, a detector, a data logger, and a display. The microphone converts the sound pressure waves into an electrical signal, amplifies it. and sends to the detector. The detector converts the electrical signal into a value that represents the SPL. The data logger stores the SPL measurements and the duration of the exposure, and the display shows the current SPL. The human body influences noise measurements, and it can change the measured values by -1 to $5\,dB$ compared to the free field measurement without a human body.

Table 2.3. SLM settings for various noise types.

Type of noise	Description	Example	Measurement time settings in SLM
Steady continuous noise	Sound level (dBA) variation during the measurement period is within 3 dB with slow settings.	Fan noise, Air compressor, large machine shop, Interior car noise, Lawn mover	Slow
Fluctuating continuous noise	Variation in sound level (dBA) during the measurement period exceeds 3 dB with slow setting.	Vacuum cleaner	Fast, max, min
Impulsive continuous noise	Sound level (dBA) variations during the measurement period exceeds the background noise by 10 dB for less than 1 sec.	Wood shop	Impulse
Steady intermittent noise	The measured SPL is less than background noise for at least two intervals in the measurement period. The sound level variation is within 3 dB of the mean.	Air grinder	Peak hold and impulse
Fluctuating intermittent noise	The measured SPL is less than background noise for at least two intervals in the measurement period. The sound level variation is more than 3 dB about the mean.	Chain saw	Fast
Impulsive intermittent noise	Measured sound has at least one impulsive noise	Shop vacuum cleaner, toy pistol	Peak hold

Dosimeters are often employed in occupational contexts where workers are regularly exposed to loud noise, such as factories or construction sites. They are also used to assess the degree of noise in a certain context, such as transit, residential, or industrial regions. Distinct standards may have distinct requirements for dosimeter measurement, such as measuring frequency range, measurement duration, mounting position, and measurement techniques. As a result, referencing the proper standard for the given measurement application is critical.

2.9 Acoustic Standard Source

A standard acoustic source is a device used as a reference for measuring sound levels or calibrating acoustic measuring equipment. These sources are designed to emit a known SPL at a specific frequency or range. They are used to test and calibrate SLMs, microphones, and other acoustic measuring equipment.

Omni-directional sound source is required for power measurement, building acoustics, and environment characterization. The common acoustic source is dodecahedron speaker which meets the ISO 3382-1 source requirement standard. The other sound sources available in the literature are impulsive in nature such as handclaps, wooden clappers, gunfire, balloons, firecrackers, laser-induced air breakdown, electric spark sources, shotshell primes, etc.

2.10 Metrics

The various metrics used in the acoustic measurement are as follows:

L_{eq} (**Equivalent sound level**): The L_{eq} is an average sound level over a specified period (T). It is a single value that describes the overall sound level during that period. An operator can select the period and frequency weighting networking. This parameter indicates the equal energy over the averaging time. The expression for L_{eq} for A-weighting networking with a period of T is given in Eq. (2.5).

$$L_{Aeq} = 10 \log_{10} \left[\frac{\frac{1}{T} \int_0^T [p_A(t)]^2 dt}{p_{\text{ref}}^2} \right] \qquad (2.5)$$

Equation (2.5) can be written for finite measurement durations as Eq. (2.6).

$$L_{Aeq} = 10 \log_{10} \left[\frac{1}{T} \sum_{i=1}^{n} t_i 10^{0.1*L_{pi}} \right] \tag{2.6}$$

Example 2.1. The measured SPLs in a workshop for three different durations are 89 dBA for 4 hours, 75 dBA for 1 hour, and 92 dBA for 4 hours. What is the L_{Aeq} for the entire 9-hour period?

Solution.

$$L_{Aeq} = 10 \log_{10} \left[\frac{1}{9} (4 * 10^{0.1*89} + 1 * 10^{0.1*75} + 4 * 10^{0.1*92}) \right]$$

$$= 90.3 \, \text{dB}$$

L_{peak} **(Peak sound level):** The L_{peak} is the highest SPL during a measurement period.

L_{min} **(Minimum sound level):** The L_{min} is the lowest SPL during a measurement period.

X-percentage exceeded level (L_x): A statistical percent exceeded level may be used if the noise level varies with time. The L_x represents the x percentage of time the level exceeds during the measurement period. For example, L_{10} means SPL which exceeds ten percent of the time, and it is used to find the "worst hour" for traffic noise. The most commonly used percentage exceeds in environmental noise measurement are:

L_1　　High amplitude sound-level events which exceed 1% of the time.

L_{10}　　High amplitude sound-level events which exceed 10% of the time.

L_{50}　　Median SPL.

L_{90}　　The level exceeded 90% of the time. It represents the background noise.

A few more parameters measured by advanced SLMs are as follows:

Frequency analyzers: These meters provide a frequency analysis of the sound level, typically in octave bands, 1/3 octave bands, and narrow bands, and are used for identifying specific frequency components that may be of concern. Band pass filters with constant percentage bandwidth are used for the octave spectral analysis. Centre frequency and bandwidth are used to specify the filters.

The use of octave band analysis is common in acoustics, particularly in measuring and analyzing environmental and industrial noise. It allows characterizing a sound based on its frequency content (the higher limit is twice the lower limit frequency of the band for octave filter). Narrow band analysis provides a way to identify and quantify specific frequency components that may be of concern. The required resolution chooses the filter type, either constant percentage or constant bandwidth.

Sound exposure level (SEL) measures the total acoustic energy received by an individual or a group over a specific period. It is typically measured in dB and is calculated by integrating the SPL over time. It is defined as

$$\text{SEL} = 10 * \log_{10} \int_0^T 10^{Lp/10} dt \tag{2.7}$$

where T is in seconds. The relation between SEL and L_{eq} is

$$\text{SEL} = L_{eq} + 10 \log_{10}(T) \tag{2.8}$$

Example 2.2. What is the SEL for an equivalent SPL of 85 dBA lasting 10 s?

Solution. By using Eq. (2.8),

$$\text{SEL} = 85 + 10 * \log 10 = 95 \, \text{dBA}$$

SEL is used to evaluate the potential for hearing damage from exposure to loud noises and the effectiveness of noise control measures. It is commonly used in occupational settings, such as in factories or construction sites, where workers are exposed to loud noise regularly.

It also evaluates a specific environment's noise level, such as transportation, residential or industrial areas.

The SEL can be calculated using a noise dosimeter or a SLM with data logging capability, which measures the SPL over a specific period. The measurement is usually done with an A-weighting filter, approximating the human ear's sensitivity to different frequencies.

It is important to note that the SEL is cumulative, meaning that exposure to loud noise for a short period can have the same impact as exposure to a lower noise level for a longer period. Limiting SEL exposure to safe levels is essential to prevent hearing damage. Following the standards, such as OSHA, MSHA, or NIOSH, is essential, which provide guidelines and limit SEL exposure.

24-hour average is calculated from the equivalent continuous levels for the 1-hour average period (L_{eq}, 1 hour). The computed 1-hour-averages averaged over 24 hours yield the 24-hour average:

$$L_{24} = 10 \log_{10} \left[\frac{1}{24} \sum_{i=1}^{24} 10^{0.1 L_{eq,1h,i}} \right] \tag{2.9}$$

Day–night average sound level (DNL) is a measure of the average sound level over 24 hours that considers both the level of the noise and the time of day when the noise occurs. It evaluates the potential for hearing damage from exposure to loud noises and the effectiveness of noise control measures.

The DNL is calculated by measuring the sound level over 24 hours, typically using a SLM or a noise dosimeter. The measurements are then adjusted to account for the fact that the human ear is more sensitive to noise at night. This adjustment is typically made by adding $10\,\mathrm{dB}$ to the sound level measurements between $10\,\mathrm{pm}$ and $7\,\mathrm{am}$.

The DNL is then calculated by averaging the adjusted sound level measurements over 24 hours. The resulting value is typically expressed in dB, and the expression is

$$L_{dn} = 10 \log_{10} \left\{ \frac{1}{24} \left[\sum_{i=8}^{22} 10^{0.1 L_{eq,1h,i}} + \sum_{i=1-7,23,24} 10^{(L_{eq,1h,i}+10)/10} \right] \right\} \tag{2.10}$$

Table 2.4.   Environmental noise measurement parameters.

Parameter	Expression	Description
Noise pollution level	$L_{NP} = L_{A,eq} + 2.56\sigma$ where $L_{A,eq}$ is the A-weighted equivalent continuous level, σ is the standard deviation of the measured noise level.	This parameter considers the variation in the measured noise level as well as the equivalent level.
Traffic noise index	$TNI = 4(L_{10} - L_{90}) + L_{90} - 30$ where L_{90} is the 90th percentile of sound level, and it accounts for background noise. L_{10} is the 10th percentile of sound level. $L_{10}-L_{90}$ incorporates the spread in the measured sound level.	This parameter represents the community reaction to traffic noise.

It is important to note that different standards may have different requirements for DNL calculation and measurement, such as the measurement frequency range, duration, and the measurement procedures, so it is vital to refer to the appropriate standard for the specific measurement application. A few other averages used in environment impact assessment are shown in Table 2.4. The list is not fully complete though.

2.11 Acoustic Environment

The measurement of acoustic pressure is straightforward as compared to the acoustic power. However, the pressure measurement depends on the environment and distance from the source. The standard acoustic environment is a free-field and diffusive field. Typically, the actual environment is not ideal; it combines both fields. Figure 2.12 shows a graph of sound level versus distance from a complex noise source. The microphone's distance from the sound source determines the boundary between the near and far fields. The free and reverberant field boundary is determined by the environment in which the measurement is performed according to the distance from the source as shown in Fig. 2.12.

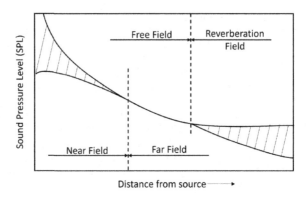

Fig. 2.12   Existence of various sound fields in interior environment as a function of distance from the source.

In the near field, sound pressure heavily depends on the proximity and the radiating characteristics of individual noise-generating mechanisms. Small changes in microphone location can show significant differences in noise level and frequency spectrum. The shaded area in the near field illustrates this. In this region, product noise emission cannot be measured.

However, microphones can be used as noise probes to aid in identifying the source of undesirable frequency components in a product noise signature. In the far field, a sound wave is propagated, and sound pressure is inversely proportional to the distance from the source. In practice, the boundary between the near field and far field is considered at about three times the largest dimension of the largest radiating surface of the sound source, and is determined by the environment in which the acoustical measurement is performed. A free field is one in which there are no reflections. This condition is often simulated in anechoic rooms to measure noise radiation and directional characteristics of sound sources. In a free field environment in the far field of a noise source, the noise level decreases by 6 dB for each doubling of distance and follows the inverse square law.

An anechoic room (or anechoic chamber) is specifically designed to be acoustically absorptive, aiming to eliminate echoes and reflections of sound waves. It is a room with walls, floor, and ceiling that absorbs sound rather than reflecting it. Anechoic rooms are constructed with materials that absorb sound waves effectively, such as fiberglass wedges, foam, or perforated metal plates. An anechoic room's walls, floor, and ceiling are typically covered with these materials to minimize sound reflections. The room is also designed to minimize sound leakage, which means the sound that comes out. It's worth noting that achieving a completely anechoic room is challenging, and it is not possible to eliminate all echoes. However, a well-designed anechoic room can reduce reflections to a low level.

Reverberant environments (diffuse fields) have sound waves impinging from all directions. The sound level theoretically is the same at all locations. These conditions are approached in a reverberation chamber. A reverberation room, also known as a reverberant chamber, is designed to create an echo or reverberation of sound. It is the opposite of an anechoic room which absorbs sound. A reverberation room is typically constructed with materials that reflect sound waves, such as concrete, marble, or metal. A reverberation room's walls, floor, and ceiling are typically covered with these materials to maximize sound reflections. It's worth noting that achieving a specific reverberation time (the time it takes for sound to decay by 60 dB) is difficult, it depends on the room's size and shape, as well as the materials used for construction, but a well-designed reverberation room can achieve a specific reverberation time.

2.12 Sound Power Measurement

Sound power measurement determines the total sound energy emitted by a source, typically measured in watts (W). It is used to quantify the total noise emitted by industrial machinery, vehicles, and other sources. A device to measure acoustic power directly is not available.

Two commonly used methods for measuring sound power are the sound intensity and the sound pressure methods.

The sound intensity method involves measuring the sound intensity at multiple points around a source using a sound intensity probe, then using these measurements to calculate the sound power by means of advanced SLM. In the far field, pressure and particle velocity are in-phase, so the intensity is a real quantity. In the near field, both the acoustic parameters (p, u) are out of phase. Therefore, intensity exists as a complex quantity, with active and reactive intensity components. The active part is the propagating energy, known as acoustic intensity. The reactive part consists of evanescent energy, which diminishes quickly as it moves away from the source. Sound intensity can be measured in virtually any environment. On-site measurements can be handled well.

Sound intensity over the test surface can be determined either by taking measurements at discrete points on the test surface or by scanning over the test surface.

Sound power is then determined as

$$W = I_n S \qquad (2.11)$$

where W is the total power radiated by the source. I_n is the surface normal intensity, and S is the area of the test surface.

The sound pressure method measures the SPL at one or more points in an enclosing surface (e.g., a hemispherical or a three-dimensional (3D) microphone array) in a standard environment such as a free field and diffuse field, and then calculating the sound power through calculations based on the measured pressure at a known distance and room constant.

The sound pressure method requires a standard environment of free field or diffusive field. The free field is established in anechoic and semi-anechoic rooms. The reverberation room is used to establish a diffuse field.

Both methods have advantages and disadvantages, and the choice of method depends on specific application and measurement conditions.

2.12.1 *Pressure method*

A more precise method of measuring the average SPL and the total power level of a source consists in making use of an anechoic room.

This is a specially constructed room to simulate a free-field environment. Its walls as well as the floor and the ceiling are lined with long thin wedges of acoustically absorbent material with power absorption coefficient of 0.99 (99%) or more in the frequency range of interest. Low-frequency noise is not absorbed easily. The lowest frequency upto which the absorption coefficient is at least 0.99 is called the cut-off frequency of the anechoic room. Such rooms are used for precise measurements as per international standards.

However, it is logistically very difficult to mount large and heavy test machines on suspended net flooring. Therefore, the Engineering Method of measurement of SPL of a machine to a reasonable accuracy is to make use of a hemi-anechoic room where one simulates a hemispherical free field. This is like testing a machine on a hard floor in an open ground. In such a room the floor is acoustically hard (highly reflective) while all four walls and the ceiling are lined with highly absorbent acoustical wedges as indicated above for an anechoic room. Coordinates of the 12 points in measurement on hypothetical hemispherical surfaces are illustrated in and listed in Fig. 2.13 and Table 2.5.

Alternatively, one can use the so-called Survey Method where measurements are made at discrete microphone locations on the parallelepiped shown in Fig. 2.14.

If S_m is the area of the hypothetical measurement surface in m^2, then

$$L_w = L_{p,av} + 10\log(S_m) \qquad (2.12)$$

For a hemispherical hypothetical surface shown in Fig. 2.13, the measurement surface area $S_m = 2\pi r^2$, and for the parallelepiped surface of Fig. 2.14,

$$S_m = 2a \times 2b + 2(2a + 2b)c = 4(ab + bc + ca), \text{ m}^2 \qquad (2.13)$$

where $2a = l_1 + 2d, 2b = l_2 + 2d, c = l_3 + d$; l_1, l_2, and l_3 are the three orthogonal dimensions of the source of noise; d is the distance (generally, 1 m) between the source and the corresponding sides of the hypothetical measurement surface, shown in Fig. 2.14.

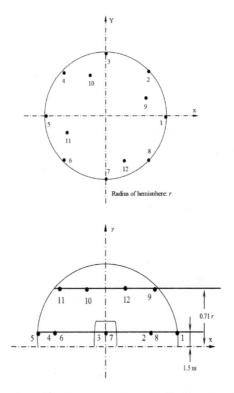

Fig. 2.13   Microphone locations on a hypothetical hemi-spherical surface.

Table 2.5.   Co-ordinates of the
12 microphones locations.

Location No.	$\dfrac{x}{r}$	$\dfrac{y}{r}$	z
1	1	0	1.5 m
2	0.7	0.7	1.5 m
3	0	1	1.5 m
4	−0.7	0.7	1.5 m
5	−1	0	1.5 m
6	−0.7	−0.7	1.5 m
7	0	−1	1.5 m
8	0.7	−0.7	1.5 m
9	0.65	0.27	0.71 r
10	−0.27	0.65	0.71 r
11	−0.65	−0.27	0.71 r
12	0.27	0.65	0.71 r

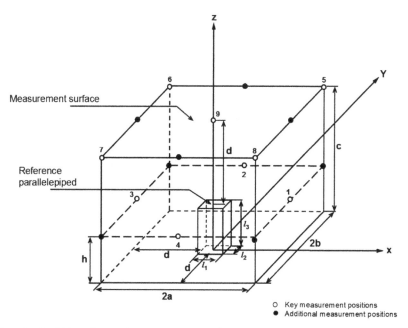

Fig. 2.14 Microphone locations on a hypothetical parallelepiped (survey method) [3].

Example 2.3. The overall dimensions of a diesel generator (DG) set are $3\,\mathrm{m} \times 1.5\,\mathrm{m} \times 1.2\,\mathrm{m}$ (height). The A-weighted SPLs at 1 m from the five $(4+1)$ radiating surfaces are 100, 95, 93, 102 and 98 dBA, respectively. Assuming that the contribution from the four walls and ceiling for the DG room is negligible, and making use of the Survey method, evaluate the sound power level of the DG set.

Solution. With reference to Fig. 2.14,

$$l_1 = 3\,\mathrm{m}, \quad l_2 = 1.5\,\mathrm{m}, \quad l_3 = 1.2\,\mathrm{m} \quad \text{and} \quad d = 1\,\mathrm{m}$$

Thus,

$$2a = 3 + 2 \times 1 = 5\,\mathrm{m}$$

$$2b = 1.5 + 2 \times 1 = 3.5\,\mathrm{m}$$

$$c = 1.2 + 1 = 2.2\,\mathrm{m}$$

Using Eq. (2.13), surface area of the hypothetical measurement surface is given by

$$S_m = 4(ab + bc + ca)$$

$$= 4\left(\frac{5}{2} \times \frac{3.5}{2} + \frac{3.5}{2} \times 2.2 + 2.2 \times \frac{5}{2}\right)$$

$$= 54.9\,\mathrm{m}^2$$

The average value of the SPL may be calculated by means of Eq. (1.44):

$$L_{p,av} = 10\log\left[\frac{1}{5}\left(10^{100/10} + 10^{95/10} + 10^{93/10} + 10^{102/10} + 10^{98/10}\right)\right]$$

$$= 98.7\,\mathrm{dBA}$$

Finally, the power level of the DG set is given by Eq. (2.12). Thus,

$$L_{WA} = 98.7 + 10\log(54.9)$$

$$= 116.1\,\mathrm{dBA}$$

2.12.2 *P–P intensity method*

Sound Intensity measurement determines the sound power radiated from the noise source. The standards for measuring sound intensity based on the pressure–pressure (P–P) method are ISO-9614/1-1993, ISO 9614/2-1996 and ISO 9614/3-2002. All these methods measure the acoustic intensity over an imaginary surface surrounding the sound-radiating surface.

A discrete-point sampling method is used in ISO 9614-1 to measure the intensity field normal to the measurement surface. The scanning-based approach is used in ISO 9614-2 and ISO 9614-3 to measure a normal intensity over the measurement surface.

Equation (2.14) describes the intensity measurement using a P–P intensity probe.

$$\hat{I}_r = \langle \hat{p}\hat{u}_r \rangle = \left\langle \frac{p_1(t) + p_2(t)}{2} \int_{-\infty}^{T} \frac{p_1(\tau) - p_2(\tau)}{\rho \Delta r} \right\rangle_t \tag{2.14}$$

where the caret (^) indicates an estimated quantity, t indicates averaging over time, ρ is the air density, and r is the microphone separation distance. Limitations that are unavoidable with this technique are due to the finite difference approximation, scattering and diffraction, the size of the probe, and instrumentation phase mismatch.

The first step in doing sound intensity measurement is to create a scanned surface surrounding the test surface. The scanned surface can be defined as a grid for guiding the intensity probe over the scanned surface.

- After preparing the grid and measuring the surface area, the sound intensity measuring probe is scanned slowly over each segment by taking two perpendicular scans. Care should be taken during scanning, so the measuring probe is always normal to the surface. Each segment records the sound intensity values. The data is then post-processed to calculate the measured surface's sound power, pressure, and intensity values.

The ideal directivity characteristic for the sound intensity probe is cosine characteristics, and it looks like a figure-of-eight pattern (in 2D). The effective spacer distance is reduced by the factor $\cos\theta$ if the sound is incident at an angle to the probe axis.

2.12.3 *Sound intensity (P–P) probe*

A sound intensity probe is a device used to measure the sound intensity in a specific location or area using microphones. It consists of two-phased matched microphones separated by a spacer for measuring sound pressure and estimating particle velocity from the pressure gradient. The signals from the two microphones are then processed by a sound intensity meter to calculate the sound intensity at that location. Figure 2.15 shows the block diagram of the pressure-based intensity measurement, and the calculation is done by means of Eq. (2.14).

Three spacer sizes, 6, 12 and 50 mm, are provided, and the one selected for use should be smaller than the smallest wavelength. For accuracy to be within 1 dB, the wavelength measured must be greater than six times the spacer distance. Thus, the 50 mm spacer can be

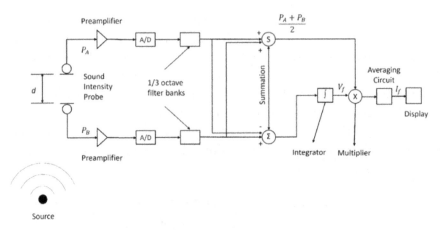

Fig. 2.15   Block diagram of the P–P sound intensity probe (courtesy Hottinger Brüel & Kjaer).

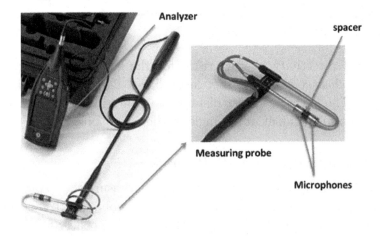

Fig. 2.16   Microphone pair with spacer (courtesy Hottinger Brüel & Kjaer).

used up to 1.25 kHz, the 12 mm spacer up to 5 kHz, and the 6 mm spacer up to 10 kHz.

Sound intensity probes as shown in Fig. 2.16 are typically used with an SLM and measurement software to perform the sound intensity measurement. The software correlates the measurements from the two microphones and calculates the sound intensity.

It is important to note that sound intensity measurements are typically made near the sound source. It is important to check the probe's calibration before and after the measurement. A few values that should be checked during intensity measurements include pressure-intensity index, pressure-residual intensity, and dynamic capability. The difference between sound pressure and sound intensity levels is defined as a pressure–intensity index ($P - I$ index). The reason for the high $P - I$ index is that the measurement environment is a diffuse field, or measurements are done perpendicular to the direction of propagation.

Pressure–residual intensity index ($p - RI$ index) is defined as the difference between the pressure and the intensity level when the same signal is fed to both channels is fixed. Ideally, intensity should be zero, but the phase mismatch causes a small phase difference between the two signals, which the analyzer interprets as intensity along the spacer.

The dynamic capability of the system is the $p - RI$ index minus the $P - I$ index. The dynamic capability should be higher than the bias error factor defined according to the Standard and accuracy requirement, as shown in Table 2.6.

2.12.4 *Acoustic particle velocity sensor*

The acoustic particle velocity sensor measures the particle velocity directly and is insensitive to the air temperature and pressure changes. It makes a highly accurate and reliable sensor for measuring acoustic particle velocity in various applications.

The working principle of the acoustic particle velocity sensor is like a hot wire anemometer. The sensor consists of several thin

Table 2.6.   Values of bias error factor and accuracy for different standards.

Bias error factor	Accuracy	Standard
7 dB	±1 dB	Survey
10 dB	±0.5 dB	Precision & Engineering

platinum strings. These strings are heated to a near-incandescent temperature. Passing sound waves alter the temperature due to convective heat transfer and, in turn, the electrical resistance of the heated strings as shown in Fig. 2.17. As the sound wave propagates, the first wire experiences a slight cooling effect, and the air absorbs some of its heat. Consequently, the heated air cools the second wire, but to a lesser extent compared to the first wire. This temperature disparity between the wires is directly related to electrical resistance change and voltage. The resulting voltage difference is proportional to the acoustic particle velocity as shown in Fig. 2.18. The particle velocity sensor is directional. Its polar pattern resembles a figure of eight. It is a bi-directional transducer.

One of the commercially available particle velocity sensors is microflown sensor. It consists of a small sensor head, typically a few millimetres in size, that contains a pair of closely spaced and precisely aligned sensors.

Both pressure and particle velocity signals are acquired through different sensors simultaneously. Such systems are sensitive to reactive sound fields. Reactivity is defined as the ratio of reactive intensity to active intensity in the log scale.

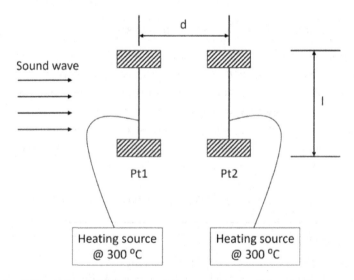

Fig. 2.17   Schematic diagram of the acoustic particle velocity sensor.

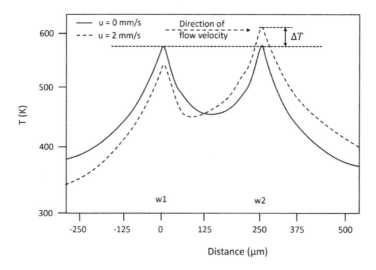

Fig. 2.18 Temperature variation across the wires for velocity variation.

2.12.5 *Reactivity*

In the $P - U$ method, the phase mismatch error between the pressure sensor and velocity sensor is more sensitive in reactive fields. Higher phase mismatch error leads to bias errors. These errors are described in terms of reactivity and reactivity error. The measured sound intensity in terms of phase error is given by [7]

$$\hat{I}_r \approx I_r \left(1 - \varphi_e \frac{J_0}{I_0}\right) \tag{2.15}$$

where $\hat{I}_r$ is measured intensity, I_0 is the "true" intensity, J_0 is the "true" reactive intensity, and φ_e phase error.

It can be observed from Eq. (2.15) that error in the measured sound intensity is directly proportional to the phase error, and it is sensitive when J_0 is high compared to I_0. The reactivity index induces uncertainty, and phase mismatch can be estimated in terms of reactivity error, and it can be expressed as Eq. (2.16) [7]:

$$\text{Error (dB)} = 10 \log_{10}(1 + \beta_e \tan \beta_f) \tag{2.16}$$

where β_e is the phase calibration error and β_f is the phase between acoustic pressure and particle velocity in the sound field.

Table 2.7.   Typical specifications of P–U probe.

Parameter	Sound pressure	Particle velocity
Diameter	12.7 mm (0.5 inch)	
Frequency range	40 Hz–8 kHz ($\pm$ 1 dB)	
	20 Hz–10 kHz ($\pm$ 2 dB)	
Sensitivity	65 mV/Pa for Sound	30 V/(m/s) @ 250 Hz
	pressure	for particle velocity
Noise floor	22 dB(A)	27 dB(A)
Temperature range	$-20°$C to 63°C	
Maximum airflow	1.5 m/s	
Connector	7 pin Lemo	

It is important to note that the Microflown sensor can be used in various measurement configurations, such as in a single point, line array, or 2D array, and they can be used in both static and dynamic measurements. They can also be used with other sensors, such as pressure transducers or particle velocity sensors, to provide a more comprehensive acoustic field measurement.

2.13 Calibrators

Calibration ensures the sensor's performance according to the specification and accuracy in the measurement instrumentation. Acoustic calibration includes reference level and frequency response.

2.13.1 *Microphone calibrator*

A microphone calibrator is a device used to calibrate a microphone for accurate sound level measurements. It generates a known SPL at a specific frequency, typically 1 kHz, and adjusts the microphone's sensitivity to match the reference level.

A microphone calibrator typically consists of a sound source, such as a loudspeaker or a signal generator, and a control unit that can adjust the level of the generated sound. The calibrator generates an

SPL that is within the range of the microphone, and the user can adjust the microphone's sensitivity to match the reference level.

There are two main types of microphone calibrators:

(1) Pistonphone calibrators: These calibrators generate a known SPL using a pistonphone, a type of sound source that generates SPL at a specific frequency of 1 kHz using a piston moving back and forth in a tube. They are widely used for calibrating microphones in the field, such as for noise measurements in industrial plants or environmental noise monitoring. The microphone size may vary, and the calibrator is coupled to the microphone to be compatible as shown in Fig. 2.19.

The user can adjust the microphone's sensitivity to match the reference level. The standard calibrator generates a 94 dB, or 114 dB SPL as shown in Fig. 2.19. The calibrator follows IEC 60942:2017 Class 1 and ANSI S1.40-2016 standards.

(2) Electric calibrators: These calibrators generate an electrical signal of known amplitude and frequency to drive the microphone amplifier circuit. The output deflection is compensated by adjusting a pre-set control. It is used for calibrating the amplifiers, filters, and weighting networks. They are typically used for laboratory measurements, such as sound system testing and studio microphone calibrating.

Fig. 2.19 Microphone calibrator, which produces an SPL of 94 dB at 1000 Hz (courtesy Hottinger Brüel & Kjaer).

2.13.2 *Sound level meter calibrator*

Sound level calibrators: These calibrators generate a known SPL using a loudspeaker and a control unit that can adjust the level of the generated sound. They are commonly used for calibrating handheld SLMs and sound level monitors. It is also worth noting that SLMs are generally calibrated to a specific standard, such as IEC 61672-1:2013 or ANSI S1.4-2006, which specifies the requirements and test methods for SLMs.

Calibration ensures that the SLM is functioning correctly and producing accurate results. Environmental factors such as temperature, humidity, and pressure can affect the sensitivity of the SLM. The most common field calibrator for SLM is the pistonphone calibrator. One of the advantages of pistonphone calibrators is that they can generate high-accuracy SPLs, typically within 0.1 dB, which is helpful in applications requiring precise measurements.

It is important to note that all acoustic calibrators should be calibrated regularly, usually with a traceable pressure standard, to ensure that the output pressure level is accurate.

2.14 Acoustic Camera

An acoustic camera, also known as a sound camera or a phased array imaging system, is a device that uses an array of microphones to capture and visualize sound waves in a specific region. These images can be used for various purposes, such as detecting and diagnosing noise sources.

It is a good tool in source ranking for steady and transient noise sources. An acoustic camera identifies and locates noise sources and analyzes objects' sound radiation patterns. It can be used in various applications, such as industrial noise control, architectural acoustics, automotive noise and vibration control, and aircraft noise control. Acoustic cameras can also be used to study structural dynamics, wind turbine noise and vibration, and bioacoustics. There are two main types of acoustical visualization techniques: acoustic beamforming and near-field acoustic holography (NAH).

Acoustic beamforming is a signal-processing technique used to control sound waves' directionality. It works by using an array of

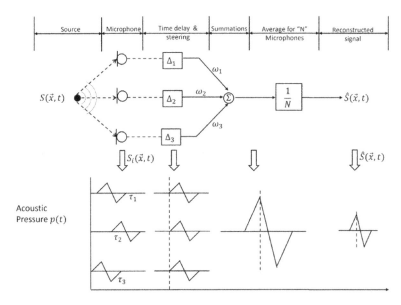

Fig. 2.20 Block diagram of beam forming (delay-sum method).

microphones to capture a sound signal and then processing the signals from each microphone to produce a "beam" of sound in a specific direction. It can be used to improve the signal-to-noise ratio of a sound signal and to reject unwanted noise or interference.

Figure 2.20 shows the block diagram of acoustic beam forming working principle. The principle behind beamforming is that the summed output from many microphones placed in an array is maximized in a specified direction and minimized in other directions. The direction of maximum response can be varied by adding an adjustable time delay to the signal from each microphone, thus "steering" the array to determine the relative intensity of sound arriving from different directions. The number and arrangement of microphones in a beamforming array influence the dynamic range, depth of focus, frequency range, and the ability to reject sound sources away from the focus point. Microphones are spaced at irregular intervals to avoid spatial aliasing and minimize the number of microphones. It can be achieved by using spiral designs.

Time delay, $\tau_i = \frac{|r_i|}{c}$, $|r_i|$ is a distance between source and microphones, c is a speed of sound, $\Delta_i = \tau_i - \min(\tau_i)$, N is number of microphones.

$$\hat{s} = \frac{1}{N} \sum_{i=1}^{N} w_i S_i(\vec{x},(t - \Delta_i)) \qquad (2.17)$$

An advantage of beamforming is that it can image distant and moving sources. However, the spatial resolution of beamformers is poor, and they do not perform well at low frequencies. For a beamforming array of largest dimension D, located at a distance L from the source, wavelength (λ), the resolution (Res), or smallest distance between two separate sources that can be resolved, is given by [12],

$$Res = 1.22 * (L/D) * \lambda$$

To ensure the entire source is mapped, the array should be placed far away from the source so that its extremities do not subtend an angle to the source extremities greater than $30°$. Ideally, the array should be placed at a distance slightly greater than or equal to one array diameter from the source.

The minimum microphone spacing limits the high-frequency useful range. Spatial aliasing limits the smallest resolved wavelength (high frequency) to double the minimum microphone spacing.

The dynamic range of a beamforming array is greatest for broadband noise sources and lowest for low-frequency tonal noise sources. It makes beamforming a useful technique for measuring mid-frequency and broadband noise.

NAH is a technique used to visualize sound sources in the near field, which is the region close to a sound source, such as an object emitting sound. NAH is an inverse array technique used to reconstruct acoustic parameters by measuring sound pressure with an array of microphones in parallel and near the sound source. The basic principle of NAH methods is described as a flow chart and shown in Fig. 2.21.

It is a form of inverse problem where the sound field is measured at many points in the near field, and then a computer algorithm is used to reconstruct the sound source distribution. NAH uses

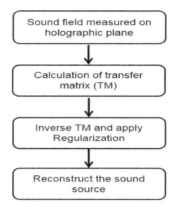

Fig. 2.21 Flowchart of the procedure involved in the NAH reconstruction techniques.

a microphone array to measure the sound field's acoustic pressure and particle velocity.

Many algorithms have been developed to reconstruct the source parameters from the measured acoustic parameters on a holographic plane. The most popular methods of NAH technique employed in the reconstruction of sound sources are [13, 14]:

- Fourier NAH (FNAH);
- Statistically Optimized NAH (SONAH);
- Equivalent Source Model (ESM);
- Inverse Boundary Element Method (IBEM).

These measured pressures are related to unknown source strength by transfer matrix. The transfer matrix (TM) element calculation varies for different NAH methods. TM can be formulated using elementary wave functions in the SONAH method. For ESM, TM can be calculated using free space Green's function, and for IBEM, TM can be obtained using Helmholtz integral equation.

As all these NAH methods are ill-posed problems, regularization is necessary to overcome the same and to reconstruct accurate results. The primary purpose is to avoid noise amplification while solving the inverse problem. Tikhonov regularization is the most used technique for NAH methods. The conjugate gradient approach is another method used in inversion. The choice of proper regularization parameter is most important to obtain better inversion results, and a few popular methods are the L-curve method, generalized cross-validation (GCV) method, and the Morozov discrepancy principle. Figure 2.21 shows the flow chart of NAH methodology for the reconstruction of noise source [12].

Each algorithm has its advantages and disadvantages, and the choice of algorithm depends on the specific application and the desired level of accuracy.

An acoustic camera comprises several hardware components that capture, process, and display acoustic data. The major components in the acoustic camera are a microphone array, DAQ system, camera, display, connectors, and pre-post processing software.

2.15 Source Identification Methods

The objective of noise source identification is to pinpoint the significant noise sources on the product based on their position, frequency characteristics, and sound power radiation. By ranking the sources, engineers can determine where design modifications have the most impact on reducing overall noise radiation. Given the multitude of available techniques, it is helpful for the NVH engineers to have an overview to aid in selecting the most suitable solution. The focus is on different methods' effective frequency range and resolution. Several methods are commonly used for noise source identification, and the choice of method depends on the specific situation and available resources. Here are a few commonly used techniques:

- **Sound level measurements:** It is a straightforward method. This method uses SLMs or microphones to map the measured SPLs at different locations. By comparing the sound levels at various positions, the general direction or area of the noise source can be determined. This method helps identify noise sources in open spaces or large areas. This method has limitations because of high measurement time, requirement of lower background noise, and suitability for the stationary noise sources only.

- **Sound intensity mapping:** Sound intensity mapping involves measuring the sound pressure and particle velocity at multiple locations using an intensity probe, as discussed in Section 2.11. Analyzing the sound intensity distribution makes it possible to identify the direction and location of the noise source with greater accuracy than simple sound level measurements because intensity is a vector quantity. This method is beneficial for identifying noise sources in complex environments or indoors. The measurement grid spacing and wavelength of the source limit the spatial resolution.

- **Beamforming:** Beamforming techniques use an array of microphones to determine the direction from which sound is coming. The noise source location and characteristics can be estimated by analyzing the time and phase differences between the signals received by different microphones. Beamforming is effective for localizing noise sources in both outdoor and indoor environments at mid and higher frequencies as discussed in Section 2.13.

- **Acoustic holography:** Acoustic holography involves capturing sound field data using an array of microphones and then reconstructing the sound field at different planes or surfaces. Analyzing the reconstructed sound field makes it possible to identify the spatial distribution of noise sources and their contributions at different frequencies. Acoustic holography is especially useful for complex noise sources and when multiple noise sources are present.

- **Sound signature analysis:** This method involves analyzing the noise signal's frequency content and temporal characteristics to identify specific patterns or signatures associated with different noise sources. By comparing the noise signature with a database of known noise sources, it is possible to determine the source type or identify the closest match. Sound signature analysis is used to identify machinery or equipment-related noise sources in industrial settings.

These methods can be used individually or in combination, depending on the noise source's complexity and the available measurement and analysis resources.

Noise source identification is an active area of research, and new techniques and technologies continue to emerge to improve the accuracy and efficiency of the process.

References

[1] Munjal, M. L., *Acoustics of Ducts and Mufflers*, Second Edition, Wiley, Chichester, UK (2014).

[2] Irwin, J. D. and Graf, E. R., *Industrial Noise and Vibration Control*, Prentice Hall, Englewood Cliffs (1979).

[3] Bies, D. A. and Hansen, C. H., *Engineering Noise Control*, Fourth Edition, Spon Press, London (2009).

[4] Acoustics — Determination of sound power levels of noise sources using sound pressure — Survey method using an enveloping measurement surface over a reflecting plane, ISO 3746: 1995(E), International Standards Organization, (1995).

[5] B&K User Manual of Hand-held Analyzer Types 2270 (2016).

[6] Bruel & Kjaer, User Manual of Sound Intensity (2016).

[7] Microflown Technologies, Scan & Paint V2.0, User's Manual (2015).

[8] Deepak, C. A., Acoustic analysis of additive manufactured multilayer periodic structures, PhD thesis, Indian Institute of Technology Hyderabad (2019).

[9] ASTM E26111-09, Standard test method for measurement of normal incidence sound transmission of acoustical materials based on the transfer matrix method 1, *Annu. B. ASTM Stand.* (2011), pp. 1–14.

[10] ASTM C423-09a, Standard test method for sound absorption and sound absorption coefficients by the reverberation room method (2009), pp. 1–12.

[11] ISO 354:2003 — Acoustics — Measurement of sound absorption in a reverberation room.

[12] Nagaraja, J., Vibro-acoustic behaviour of flexible rectangular ducts, PhD thesis, Indian Institute of Technology Hyderabad (2018).

[13] Bai, M. R., Ih, J. G., and Benesty, J., *Acoustic Array Systems: Theory, Implementation, and Application*, John Wiley & Sons, Singapore (2013).

[14] Williams, E. G., *Fourier Acoustics: Sound Radiation and Near-field Acoustical Holography*, Academic Press, Cambridge, UK (1999).

Problems

2.1. Describe the type 1 and type 2 SLMs.

2.2. Describe frequency weighting curves.

2.3. Describe time weighting and its uses for various noise source types.

2.4. How is the SLM calibrated?

2.5. Define free field and diffuse field with an example?

2.6. Explain the far field and near filed in an interior environment.

2.7. Explain the different power measurement methods based on pressure and intensity.

2.8. Which type of microphone is best for precision measurement like sound Intensity? Explain the working principle of that microphone?

2.9. Sound intensity measuring probe has two microphones separated with a spacer. Discuss the effect of the spacer on measuring frequency.

2.10. The technical specification sheet of a condenser microphone has been misplaced. In order to measure sensitivity of the microphone, it is subjected to SPL of 90 dBA in the 1000 Hz octave band. The output voltage is measured to be 1.0 mV. Estimate the sensitivity of the microphone.

[Ans.: −56 dB re 1 V/Pa]

2.11. An omni-directional dynamic microphone open-circuit is specified as −80 dB for 150 Ω case. It is also specified that 0 dB = 1 V/μ bar. What would be the open-circuit voltage in V (in volts)?

[Ans.: 0.001 V]

2.12. To measure sensitivity of the microphone, it is subjected to SPL of 94 dB at 1 kHz. The output voltage of microphone is measured to be 1.585 mV. Find the sensitivity of the microphone.

[Ans.: −56 dB re 1 V/Pa]

2.13. Pink noise is measured in the frequency range of 50–5000 Hz in 1/3 octave band frequency range using a Class I SLM which has a microphone sensitivity of −25 dB ref 1.0 V/Pa. The total measured voltage is 0.05 V.

(a) Calculate the sound pressure spectrum in 1/3 octave band frequency.

(b) Calculate the sound pressure spectrum in octave band frequency.

**[Ans.: (a) Band value, 79.8 dB, Total 93 dB,
(b) Band value, 84.6 dB]**

2.14. A worker operates near a noisy machine in a workshop. While the machine idles, it produces a level of 85 dB(A). When it operates, it produces a level of 90 dB(A) at work position.

(i) If the machine runs only 10% of the shift hours (8 hours), compute the A-weighted, 8-hour equivalent noise level.
(ii) How much should the exposure be reduced (in hours) to comply with the 80 dB(A) criterion?

2.15. If the sound intensity of machine is $0.01 \, \text{W/m}^2$ at 2.5 m away, what would be sound intensity level at 7.5 m?

[Ans.: 90.5 dB]

2.16. In a particular community, the daytime L_{eq} was 77 dBA and the nighttime L_{eq} was 58 dBA. Determine the day-night A-weighted average sound level?

[Ans.: 75.3 dBA]

2.17. A forging hammer operation cycle rate is 72 cycles per minute, and the equivalent SPL for a minute duration is 89 dBA. What is its SEL?

[Ans.: 106.8 dBA]

2.18. Determine the error in the intensity meter reading if the microphone spacing is 12 mm. The frequency of the sound wave is 8 kHz, and the speed of sound in the air around the microphone is 340 m/s. Derive the necessary equations to estimate the error in the intensity meter.

[Ans.: 12.61%]

2.19. SPL variation with respect to distance from source in an enclosed space is given in the table below. Describe the existence of different sound fields and calculate the critical distance from the source?

Distance (m)	0.61	1.22	2.44	3.05	4.57	6.1	9.14	15.24	21.34	30.48
L_p(dB)	112	100	94	92	88.5	87	86	84	85	85

[**Ans.: Up to 1.22 m, Near field, 4.57 m, Free field, 6.1 m, Diffuse field, Critical distance**]

2.20. A quieter manufacturing machine requires a tool change operation four times during its 8-h operation (working day). During one such operation, the measured equivalent SPL for 1-min measurement is 90 dB.

(a) Calculate the sound exposure level for one event.

(b) Calculate the operator's personal exposure level for a working day.

[**Ans.: 107.8 dB, 69.2 dB**]

Chapter 3

Vibration and Its Measurement

3.1 Introduction

Vibration is an oscillatory motion of a particle or an object about its mean position. This motion may or may not be periodic. When vibration is caused by the unbalanced forces or moments of a reciprocating or rotary machine rotating at a constant speed, then it is periodic. In this chapter, and indeed in this textbook, vibration and the resultant acoustic pressure are assumed to be periodic.

Vibration is caused by the interaction of two primary elements: mass and spring (stiffness), or in other words, inertia and elasticity. Damping represents a third element, which is invariably there in the system but is not an essential or primary element for vibration to take place. Mass and spring are characterized by kinetic energy and potential energy, respectively. During a steady state oscillation, there is a continuous exchange between the two types of energy, with their total remaining constant. The external excitation, then, supplies just enough energy to compensate for the energy dissipated into heat by the damping in the system, or radiated out as acoustic energy. In the absence of external excitation, the system would execute free vibrations (if initially disturbed and left alone) with successively decreasing amplitude depending on the amount of damping in the system and radiation resistance on the vibrating surfaces.

In this book, vibration and its control have been dealt with primarily insofar as vibrating structures radiate noise. However, excessive vibration can degrade the performance and decrease the fatigue life of machine bearings and structures, resulting in economic loss. In extreme cases, they may cause fatal accidents.

Study of vibrations involves Newton's laws of motion and hence the science of Dynamics. Vibratory systems are therefore called dynamical systems. These may be classified as linear or nonlinear, lumped or distributed, conservative or nonconservative, depending upon the governing differential equations, and the absence or presence of damping, respectively.

The number of independent coordinates required to describe motion completely is termed the degrees of freedom (DOF) of the system. All basic features of vibration of a multi-degree of freedom (MDOF) dynamical system may be easily understood by means of a single-degree-of freedom (SDOF) system as follows.

3.2 Vibration of a Single Degree of Freedom System

For a linear lumped-parameter SDOF system shown in Fig. 3.1, making use of the free body diagram and the Newton's Second Law of Motion, the instantaneous displacement $x(t)$ of the lumped mass m is given by the equation,

$$m\ddot{x}(t) = f(t) - kx(t) - c\dot{x}(t)$$

or

$$m\ddot{x}(t) + c\dot{x}(t) + kx(t) = f(t) \qquad (3.1)$$

where m is the mass, c is the damping coefficient, k is the stiffness of the spring, $f(t)$ is the excitation or external force acting on the

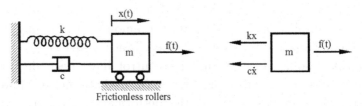

Fig. 3.1 A SDOF system with viscous damping along with the free-body diagram of the mass, m.

mass, $\dot{x}(t) \equiv dx/dt$ is the instantaneous velocity, and $\ddot{x}(t) \equiv d^2x/dt^2$ is the instantaneous acceleration of the mass in the positive (left to right) direction.

Equation (3.1) is a linear, second-order, ordinary differential equation with constant coefficients. Its solution consists of a complementary function and particular integral, representing the free vibration and forced vibration responses of the system, respectively.

3.2.1 *Free vibration*

The free vibration response of an underdamped system may be seen to be [1]:

$$x(t) = e^{-\zeta \omega_n t}(A \sin \omega_d t + B \cos \omega_d t) \tag{3.2a}$$

where $\omega_n = (k/m)^{1/2}$ is the natural frequency (in rad/s) of the undamped system (when $c = 0$), $\omega_d = \omega_n \left(1 - \zeta^2\right)^{1/2}$ is the natural frequency of the damped system, $\zeta = c/c_c$ is the damping ratio of the system, and $c_c = 2m\omega_n = 2(km)^{1/2}$ is the critical damping of the system, beyond which there would be no oscillation; the mass would approach the mean position asymptotically.

A and B are arbitrary constants that can easily be determined from the initial displacement x_0 and velocity u_0 of the mass, m. Thus, it can readily be seen that,

$$B = x_0 \quad \text{and} \quad A = \frac{u_0}{\omega_d} + \frac{\zeta \omega_n}{\omega_d} x_0 = \frac{u_0}{\omega_d} + \frac{\zeta}{(1 - \zeta^2)^{1/2}} x_0 \tag{3.2b}$$

For most mechanical vibro-acoustic systems, the damping ratio is much less than unity, or in other words, the damping coefficient is much less than the critical damping. Thus,

$$\zeta \ll 1 \quad \text{or} \quad c \ll c_c, \quad \text{and} \quad \omega_d \simeq \omega_n \tag{3.3}$$

Figure 3.2 shows the typical free vibration response of such a system, where mass m is given an initial displacement x_0, and released with zero initial velocity.

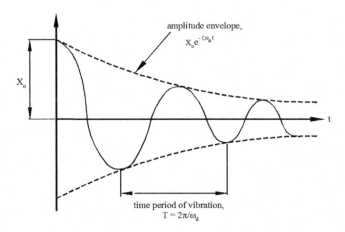

Fig. 3.2   Free vibration of a typical lightly damped system.

The undamped natural frequency in Hertz, or cycles per second, and the corresponding time period in seconds are given by

$$f_n = \frac{\omega_n}{2\pi}, T = \frac{1}{f_n} = \frac{2\pi}{\omega_n} \qquad (3.4)$$

Example 3.1. The ratio of two successive amplitudes of the damped spring-mass system of the type shown in Fig. 3.1, when disturbed and left to vibrate freely, is 0.9. If the mass is 10 kg and the spring stiffness is 10 N/mm, evaluate the damping ratio, critical damping and the damped natural frequency of the system.

Solution.

Mass, $m = 10\,\text{kg}$

Stiffness, $k = 10\frac{N}{mm} * \frac{1000\,\text{mm}}{1\,\text{m}} = 10000\,\text{N/m}$

Undamped natural frequency of the system,

$$\omega_n = \left(\frac{k}{m}\right)^{1/2} = \left(\frac{10000}{10}\right)^{1/2} = 31.62\,\text{rad/s}$$

As per Eq. (3.2a) and Fig. 3.2, the ratio of successive amplitudes is given by,

$$\frac{X_2}{X_1} = e^{-\zeta\omega_n T}, T = 2\pi/\omega_d, \omega_d = \omega_n(1 - \zeta^2)^{1/2}$$

Thus,

$$e^{-2\pi\zeta/(1-\zeta^2)^{1/2}} = 0.9$$

or

$$\frac{2\pi\zeta}{(1-\zeta^2)^{1/2}} = \ln\left(\frac{1}{0.9}\right) = 0.105$$

or

$$\zeta^2 = \left(\frac{0.105}{2\pi}\right)^2 (1-\zeta^2) = 0.00028(1-\zeta^2)$$

Thus, damping ratio of the system is given by

$$\zeta = \left(\frac{0.00028}{1+0.00028}\right)^{1/2} = 0.0167$$

Critical damping of the system, $c_c = 2m\omega_n$

$$= 2^*10^*31.62$$

$$= 632.4\,\mathrm{Ns/m}$$

Finally, damped natural frequency of the system,

$$\omega_d = \omega_n(1-\zeta^2)^{1/2}$$

$$= 31.62\{1-(0.0167)^2\}^{1/2}$$

$$= 31.62^*0.9998 = 31.62\,\mathrm{rad/s}$$

$$= \frac{31.62}{2\pi} = 5.03\,\mathrm{Hz}$$

This natural frequency and the associated free vibration parameters play a seminal role in our understanding of the forced response of the system, which is of primary concern in the field of noise and vibration control.

3.2.2 *Forced response*

A periodic forcing function with time period T may be expressed as a Fourier series.

$$f(t) = a_0 + \sum_{n=1}^{\infty} a_n \cos(n\omega_0 t) + b_n \sin(n\omega_0 t), \omega_0 = \frac{2\pi}{T} \qquad (3.5)$$

or as an exponential series

$$f(t) = \sum_{n=-\infty}^{\infty} c_n e^{jn\omega_0 t} \qquad (3.6)$$

where,

$$a_0 = \frac{1}{T} \int_0^T f(t) dt$$

$$a_n = \frac{2}{T} \int_0^T f(t) \cos(n\omega_0 t) dt \qquad (3.7)$$

$$b_n = \frac{2}{T} \int_0^T f(t) \sin(n\omega_0 t) dt$$

$$c_n = (a_n - jb_n)/2 = \frac{1}{T} \int_0^T f(t) e^{-jn\omega_0 t} dt$$

In this book, the exponential form (Eq. (3.6)) will be used, with

$$\omega = n\omega_0 \qquad (3.8)$$

For a linear system, the principle of superposition holds; i.e., response of a periodic forcing function may be expressed as sum of the system responses to individual harmonics. Thus, Eq. (3.1) for the particular integral may be written as:

$$m\ddot{x} + c\dot{x} + kx = Fe^{j\omega t} \qquad (3.9)$$

Obviously, the steady state response to the harmonic forcing function is given by

$$x(t) = \frac{Fe^{j\omega t}}{-m\omega^2 + j\omega c + k} \equiv Xe^{j\omega t} \qquad (3.10)$$

where the complex amplitude of the steady state displacement is given by

$$X = \frac{F}{(k - m\omega^2) + j(\omega c)} \qquad (3.11)$$

This may be rewritten in the non-dimensional form [2, 3]:

$$\frac{X}{F/k} = \frac{1}{1 - \left(\frac{\omega}{\omega_n}\right)^2 + j2\zeta\frac{\omega}{\omega_n}} \qquad (3.12)$$

or

$$\frac{|X|}{x_{st}} = \frac{1}{\{(1 - r^2)^2 + (2\zeta r)^2\}^{1/2}} \qquad (3.13)$$

where $r = \omega/\omega_n$ is the frequency ratio, and $x_{st} = F/k$ is the static displacement which serves as a normalization factor.

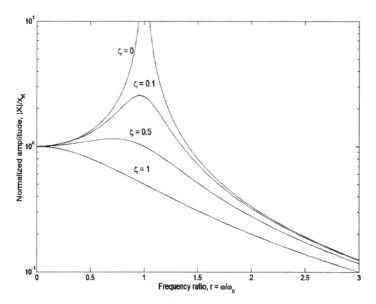

Fig. 3.3 Normalized displacement of mass of the single DOF system of Fig. 3.1 with damping ratio as parameter.

The non-dimensional forced response or steady-state response given by Eq. (3.12) is plotted in Fig. 3.3, whence the following observations may be made:

(a) At very low frequencies, ($r \ll 1$) the dynamic response amplitude $|X|$ tends to the static x_{st} displacement.

(b) In the absence of damping, the system response would tend to infinity when the forcing frequency ω approaches the undamped natural frequency ω_n. This phenomenon is termed Resonance. In fact, the spring of Fig. 3.1 would break in no time if the system were excited at its natural frequency.

(c) At higher frequencies ($\omega \gg \omega_n$), the dynamic response amplitude decreases monotonically as

$$\frac{|X|}{x_{st}} \simeq \left(\frac{\omega_n}{\omega}\right)^2 \tag{3.13a}$$

Thus, it is imperative to design the system with as low a natural frequency as feasible.

(d) The restraining effect of damping is limited to the resonance frequency and its immediate neighborhood. Nevertheless, its role is crucial.

(e) In a damped single DOF system, resonance peak does not occur precisely at $r = 1$ or $\omega = \omega_n$. As can be seen from Eq. (3.13) and Fig. 3.3, it occurs at a slightly lower frequency. However, for lightly damped system ($c \ll c_c$ or $\zeta \ll 1$), this shift can be neglected.

Example 3.2. A machine of 500 kg mass when lowered onto a set of springs causes a static displacement of 1.0 mm. Evaluate (a) overall stiffness of the springs, and (b) natural frequency of the spring mass system (assume the foundation to be rigid).

Solution.
Static displacement, $x_{st} = \frac{mg}{k}$

whence,

$$\text{stiffness, } k = \frac{mg}{x_{st}} = \frac{500 * 9.81}{1/1000} = 4905 \, \text{kN/m}$$

Natural frequency, $f_n = \frac{\omega_n}{2\pi} = \frac{(k/m)^{1/2}}{2\pi} = \frac{(4905 \times 10^3 / 500)^{1/2}}{2\pi} =$ 15.76 Hz.

3.3 Vibration of a Multiple Degrees of Freedom System

A dynamical system with more than one lumped masses interconnected with spring (and damper) elements would require more than one coordinates for description of the instantaneous displacements (and velocities). Such a system is said to have, or can be characterized by, MDOF. An example of an undamped three DOF system is shown in Fig. 3.4, where, for convenience, the forcing function is assumed to be harmonic.

Equations of dynamical equilibrium for the system of Fig. 3.4 are given by the matrix equation [1]:

$$\begin{bmatrix} m_1 & 0 & 0 \\ 0 & m_2 & 0 \\ 0 & 0 & m_3 \end{bmatrix} \begin{bmatrix} \ddot{x}_1 \\ \ddot{x}_2 \\ \ddot{x}_3 \end{bmatrix} + \begin{bmatrix} k_1 & -k_1 & 0 \\ -k_1 & k_1 + k_2 & -k_2 \\ 0 & -k_2 & k_2 + k_3 \end{bmatrix} \begin{bmatrix} x_1 \\ x_2 \\ x_3 \end{bmatrix} = \begin{bmatrix} F_0 e^{j\omega t} \\ 0 \\ 0 \end{bmatrix}$$

$$(3.14)$$

As three coordinates (x_1, x_2, x_3) are sufficient to characterize the system of Fig. 3.4, it is three-DOF system. Equation (3.14) may be re-written in the following matrix form:

$$[M]\{\ddot{x}\} + [K]\{x\} = \{F\}e^{j\omega t} \qquad (3.15)$$

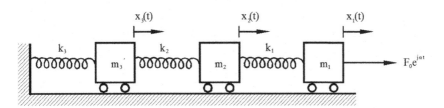

Fig. 3.4   Example of a 3-DOF dynamical system.

where

$\{x\}$ is the displacement vector $[x_1, x_2, x_3]^T$
$\{\ddot{x}\}$ is the corresponding acceleration vector $[\ddot{x}_1, \ddot{x}_2, \ddot{x}_3]^T$
$[M], [K]$ and $\{F\}$ are the inertia (or mass) matrix, stiffness matrix, and force vector, respectively.

3.3.1 *Free response*

Free vibration of the system is governed by the equation,

$$[M]\{\ddot{x}\} + [K]\{x\} = \{0\} \tag{3.16}$$

where $\{0\}$ is a null vector $\begin{bmatrix} 0 & 0 & 0 \end{bmatrix}^T$, representing free vibration. Substituting $\{x(t)\} = \{X\}e^{j\omega t}$ in Eq. (3.16), where ω is a natural frequency, we get a set of three (in general, n) homogeneous equations which can be arranged in the matrix form:

$$([K] - \omega^2[M])\{X\} = \{0\} \tag{3.17}$$

For these equations to be consistent, the determinant of the coefficient matrix must be zero. This condition yields the frequency equation of the system:

$$\left| [K] - \omega^2[M] \right| = 0. \tag{3.18}$$

Roots of this equation $(\omega^2 = \omega_1^2, \omega_2^2, \ldots, \omega_n^2)$ represent the square of the natural frequencies of the n-DOF system. For the system shown in Fig. 3.4, $n = 3$, and therefore this particular system will have three discrete natural frequencies.

3.3.2 *Forced response of a multi-DOF system*

Forced response or steady-state response or dynamic response of a multi-DOF system can similarly be evaluated from the inhomogeneous matrix equation (3.15):

$$\{x(t)\} = ([K] - \omega^2[M])^{-1}\{F\}e^{j\omega t} \tag{3.19}$$

Here, ω is the forcing frequency. It is assumed in Eq. (3.19) that each mass is excited harmonically with the same frequency ω.

Formal similarity of Eq. (3.19) with Eq. (3.10) may be noted. Computationally efficient and inherently stable algorithms, and the corresponding function sub-programs, are available for inversion of the dynamic matrix, $[H]$ in Eq. (3.19) above:

$$[H] \equiv [K] - \omega^2 [M] \tag{3.20}$$

As indicated by Eqs. (3.14) and (3.20), the inertia matrix $[M]$, stiffness matrix $[K]$, and hence the dynamic matrix $[H]$ are symmetric for linear, passive reciprocal systems. Often, $[H]$ is tridiagonal or banded matrix. Therefore, inversion of $[H]$ is a relatively simple and fast operation.

3.3.3 *Modal expansion*

When an n-DOF system vibrates at one of its natural frequencies, all of its masses (or lumped inertias) vibrate in phase and the vector of their relative amplitudes represents an eigen vector or modal vector, $\{u\}$. When the modal vectors corresponding to all natural frequencies are arranged column wise, they constitute an $n \times n$ modal matrix.

Natural frequencies (or eigenvalues) and the modal matrix (or eigenmatrix) can be computed simultaneously by means of Eigenvalue analysis making use of one of the several algorithms available in Linear Algebra. Computationally efficient subroutines or Function subprograms are available in the FORTRAN or MATLAB function libraries.

The normal modes or vectors are found to be orthogonal to each other in the following sense [2]:

$$\{u_i\}^T [M] \{u_j\} = \{0\}, \{u_i\}^T [K] \{u_j\} = \{0\} \quad \text{for } i \neq j \tag{3.21}$$

As a modal vector represents relative amplitudes of different masses during free vibration of the system at one of its natural frequencies, it is desirable to normalize modal vectors such that,

$$\{u_i\}^T [M] \{u_i\} = I, \quad i = 1, 2, \ldots, n \tag{3.22}$$

Orthogonality relations (3.21) may be used to decouple the equations of motion of a multi-DOF system. Thus, the solution of the original coupled equations may be reduced to the solution of n-independent differential equations. This method of solution is called the Modal Expansion or Eigenfunction Expansion method, and is particularly useful for the evaluation of the response of a multi-DOF dynamical system to arbitrary input, $f(t)$. The response or solution may, then, be expressed as a series of the system eigenvectors [2]:

$$\{x(t)\} = \sum_{i=1}^{n} q_i(t)\{u_i\} \tag{3.23}$$

Here, the coefficients $q_i(t)$ are called Modal Coordinates, and represent a real transformation from the physical coordinates. These may be determined by substituting Eq. (3.23) into the original governing equation,

$$[M]\{\ddot{x}\} + [K]\{x\} = \{f(t)\}, \tag{3.24}$$

and making use of the orthogonality relations (3.21) and normalization relations (3.22). Thus, one obtains the uncoupled or independent differential equations.

$$\{\ddot{q}_i(t)\} + \omega_i^2\{q_i(t)\} = \{u_i\}^T\{f(t)\}, i = 1, 2, \ldots, n \tag{3.25}$$

which can be solved easily. Substituting these modal coordinates in Eq. (3.23), one obtains a closed-form modal expansion of the total response, $x(t)$.

It may be noted that $\{u_i\}^T\{f(t)\}$ on the RHS of Eq. (3.25) represents a modal decomposition of the input force vector inasmuch as it represents the component of the input vector that would excite only the ith mode.

Now, if the forcing vector were harmonic, i.e., if,

$$\{f(t)\} = \{F\}e^{j\omega t} \tag{3.26}$$

then, the solution of Eq. (3.25) would become,

$$q_i(t) = \frac{\{u_i\}^T\{F\}}{\omega_i^2 - \omega^2}e^{j\omega t} \tag{3.27}$$

Obviously, if the forcing frequency ω were equal to ω_i, the natural frequency of the ith mode of vibration of the MDOF system, then $q_i(t)$, amplitude of the ith mode, would build up until the system broke up, unless the system had substantial inherent damping. This underlines the need for designing-in sufficient damping in every physical dynamical system (like an automobile) that is going to be subjected to a periodic forcing function as well as random excitation. Some practical ways of effecting this are discussed later in Chapter 4.

3.4 Transmissibility

The unbalanced forces and moments of a reciprocating machinery (like a reciprocating compressor or engine) or rotating machinery (like a turbine) not only make the machine vibrate but are also transmitted in part to the foundation. The oscillating forces so transmitted propagate through the floor as structure borne sound and radiate audible sound elsewhere. Similarly, oscillation of a surface supporting a sensitive instrument may transmit motion to the instrument. These two phenomena are shown in Fig. 3.5.

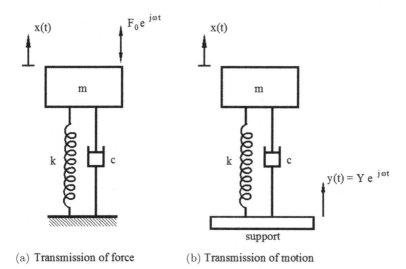

(a) Transmission of force   (b) Transmission of motion

Fig. 3.5   Transmissibility of force and motion.

Equation of motion of mass m in the single DOF damped system of Fig. 3.5(a) is given by

$$m\ddot{x} + c\dot{x} + kx = F_0 e^{j\omega t} \tag{3.28}$$

whence

$$x(t) = \frac{F_0}{k - m\omega^2 + j\omega c} e^{j\omega t} \tag{3.29}$$

The force transmitted to the foundation is

$$F_T = c\dot{x} + kx = (j\omega c + k)x(t) = \frac{k + j\omega c}{k - m\omega^2 + j\omega c} F_0 e^{j\omega t} \tag{3.30}$$

and then, transmissibility TR is given by

$$TR \equiv \frac{|F_T|}{F_0} = \frac{(k^2 + \omega^2 c^2)^{1/2}}{\{(k - m\omega^2)^2 + \omega^2 c^2\}^{1/2}} \tag{3.31}$$

Now, referring to the single DOF damped system of Fig. 3.5(b), the equation of motion is given by

$$m\ddot{x} + c(\dot{x} - \dot{y}) + k(x - y) = 0$$

which can be rearranged as

$$m\ddot{x} + c\dot{x} + kx = c\dot{y} + ky$$

For harmonic excitation $y(t) = Ye^{j\omega t}$, the response will also be harmonic: $x(t) = Xe^{j\omega t}$, and then the motion transmissibility is given by

$$TR \equiv \frac{|X|}{Y} = \left| \frac{j\omega c + k}{-m\omega^2 + j\omega c + k} \right| = \frac{(k^2 + \omega^2 c^2)^{1/2}}{\{(k - m\omega^2)^2 + \omega^2 c^2\}^{1/2}} \tag{3.32}$$

It may be noted that Eq. (3.32) for the motion transmissibility is identically similar to Eq. (3.31) for the force transmissibility.

This common expression for transmissibility has the following non-dimensional form [4]:

$$TR = \left[\frac{1 + (2\zeta r)^2}{(1 - r^2)^2 + (2\zeta r)^2} \right]^{1/2} \tag{3.33}$$

where, as defined earlier in Section 3.2.1, $r = \omega/\omega_n$ is the frequency ratio, and $\zeta = c/c_c$ is the damping ratio.

Equation (3.33) is plotted in Fig. 3.6. As r tends to unity, TR is inversely proportional to damping ratio ζ. Thus, at around the resonance frequency, in the absence of any damping, TR would tend to infinity. Thus, damping plays a crucial role at and around the resonance frequency ($\omega \simeq \omega_n$). However, it is somewhat counter-productive at $\omega > \sqrt{2}\omega_n$, as may be observed from Fig. 3.6.

Figure 3.6 also shows that for the force or motion transmissibility to be much less than one, the frequency ratio must be much more

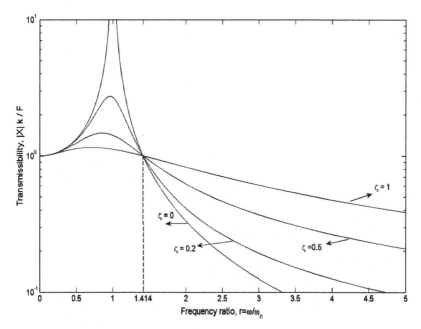

Fig. 3.6 Transmissibility of a single DOF system with damping ratio as a parameter.

than unity. For a lightly damped system with $\omega \gg \omega_n$,

$$\text{TR} \approx \frac{(1 + 4\zeta^2 r^2)^{1/2}}{r^2} \quad \text{for } r \gg 1 \tag{3.34}$$

This equation may be observed to be similar to Eq. (3.13a) for the response of mass m in Fig. 3.5(a):

$$\frac{X}{F_0/k} \approx \frac{1}{r^2} \quad \text{for } r \gg 1 \tag{3.34a}$$

It follows from Eqs. (3.34) and (3.34a) that in order to reduce the vibration of the machine as well as to reduce the unbalanced forces transmitted to the foundation, the spring stiffness should be as small as feasible so that the natural frequency of the system is much lower than the excitation frequency. This requirement can also be met by means of an inertia block shown in Fig. 3.7.

Generally, the inertia block mass m_2 is several times the machine mass m_1. For rigid foundation, natural frequency of the dynamical

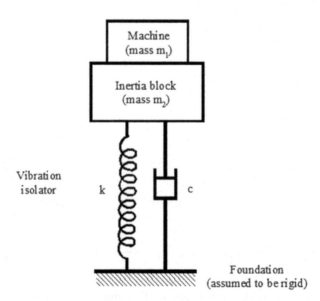

Fig. 3.7   Use of an inertia block.

system of Fig. 3.7 is given by

$$\omega_n = \left(\frac{k}{m_1 + m_2}\right)^{1/2} \qquad (3.35)$$

Example 3.3. A reciprocating compressor of mass 100 kg is supported on a rigid foundation through a spring of stiffness 1000 N/cm in parallel with a dashpot of damping coefficient 100 Ns/m. If it is acted upon by a vertical unbalanced force of 1000 N amplitude at a frequency corresponding to 1500 RPM, then evaluate amplitude of (a) vibration of the compressor body, and (b) the force transmitted to the rigid foundation.

Solution.

Mass, $m = 100$ kg (given).

Stiffness, $k = 1000\frac{\text{N}}{\text{cm}} = 1000\frac{\text{N}}{0.01\,\text{m}} = 10^5$ N/m

Damping coefficient, $c = 100$ Ns/m (given)

Forcing frequency, $\omega = \frac{2\pi\text{RPM}}{60} = \frac{2\pi \times 1500}{60} = 157.1$ rad/s

Undamped natural frequency,

$$\omega_n = \left(\frac{k}{m}\right)^{1/2} = \left(\frac{10^5}{100}\right)^{1/2} = 31.62\,\text{rad/s}$$

Frequency ratio, $r = \frac{\omega}{\omega_n} = \frac{157.1}{31.62} = 4.97$

Amplitude of the excitation force, $F_0 = 1000$ N (given).

(a) Amplitude of vibration of the compressor body is given by Eq. (3.29):

$$X = \frac{F_0}{\{(k - m\omega^2)^2 + \omega^2 c^2\}^{1/2}}$$

$$= \frac{1000}{[\{10^5 - 100 \times (157.1)^2\}^2 + (157.1 \times 100)^2]^{1/2}}$$

$$= \frac{1000}{2.37 \times 10^6} = 0.42\,\text{mm}$$

(b) Amplitude of the unbalanced force transmitted to the rigid foundation is given by Eq. (3.30):

$$F_T = \frac{(k^2 + \omega^2 c^2)^{1/2}}{\{(k - m\omega^2)^2 + \omega^2 c^2\}^{1/2}} F_0$$

$$= \frac{\{(10^5)^2 + (157.1*100)^2\}^{1/2}*1000}{[\{10^5 - 100(157.1)^2\}^2 + (157.1*100)^2]^{1/2}}$$

$$= \frac{1.012*10^8}{2.37*10^6} = 42.7\,\text{N}$$

Incidentally, force transmissibility,

$$TR = \frac{F_T}{F_0} = \frac{42.7}{1000} = 0.0427 = 4.27\%$$

It is worth noting that the vibration isolation system of Example 3.3 is very efficient in as much as amplitude of vibration of the compressor body is as small as 0.42 mm and the force transmissibility is a meager 4.27%. This effectiveness is due to the fact that the natural frequency is so small that frequency ratio,

$$r = \frac{\omega}{\omega_n} = \frac{157.1}{31.62} = 4.97$$

which is much more than unity. This is indeed the basic principle underlying the design of vibration isolator of a single DOF system.

3.5 Critical Speed

Critical speed is the angular speed of rotation of a rotor which coincides with the flexural natural frequency of the rotor-shaft-bearings system.

Let a rotor or disc of mass m be mounted in the middle of a shaft of length l, as shown in Fig. 3.8. There would invariably be a little unbalance (eccentricity of the centre of gravity) in the rotor, resulting in a centrifugal force $m\Omega^2 e$, where e is the eccentricity and Ω is the angular speed of rotation of the system. It will have a vertical component $m\Omega^2 e \sin\theta$ where $\theta = \Omega t$. This will result in a vertical

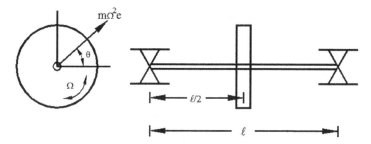

Fig. 3.8   A rotor and shaft mounted on simple support bearings.

oscillation of the rotor, $y(t)$, governed by the equation:

$$m\ddot{y} + k_f y = m\Omega^2 e \sin(\Omega t) \qquad (3.36)$$

where k_f is the flexural stiffness of the simply supported shaft [5], $k_f = 48\text{EI}/1^3, EI$ is the flexural rigidity of the shaft. Here, mass of the shaft and damping of the bearings have been neglected for simplicity. Equation (3.36) has the following steady-state solution:

$$y(t) = \frac{m\Omega^2 e}{k_f - m\Omega^2} \sin(\Omega t) \qquad (3.37)$$

It can be rearranged in the non-dimensional form,

$$y(t) = \frac{r_f^2}{1 - r_f^2} e \cdot \sin(\Omega t) \qquad (3.38)$$

where $r_f = \frac{\Omega}{\omega_{n,f}}$ is the flexural frequency ratio, $\omega_{n,f} = (k_f/m)^{1/2}$ is the flexural natural frequency for the simplified single DOF system.

Clearly, $y(t)$ would tend to infinity if r_f were to be unity, or if $\Omega = \omega_{n,f}$. Therefore, the critical speed of rotation of the rotor is $\omega_{n,f}$ rad/s or $(\omega_{n,f}/2\pi) \times 60$ revolutions per minute. Hence, the critical speed N_c is given by

$$N_c = \frac{60}{2\pi} \left(\frac{48\text{EI}}{ml^3} \right)^{1/2}, \text{RPM} \qquad (3.39)$$

This is of course an approximate estimate of the critical speed of the simplified or idealized undamped single DOF rotor shown in Fig. 3.8. Nevertheless, it explains the concept of critical speed.

Example 3.4. A steel disc of diameter 0.5 m and width 5 cm is mounted in the middle of a 2 cm diameter, 1 m long shaft mounted on ball bearings at the two ends. Evaluate the critical speed of the rotary system.

Solution.

Diameter of the rotary disc, $D = 0.5$ m (given).
Axial width (or length) of the disc, $b = 0.05$ m (given).
For steel, elastic modulus, $E = 2{*}10^{11}$ N/m^2
density, $\rho = 7800$ kg/m^3
Mass of the disc, $m = \rho \frac{\pi D^2}{4} b$

$$= \frac{7800^*\pi(0.5)^{2*}0.05}{4}$$

$$= 76.58 \text{ kg}$$

Diameter of the shaft, $d = 0.02$ m (given).
Moment of inertia of the shaft, $I = \frac{\pi d^4}{64} = \frac{\pi^*(0.02)^4}{64}$

$$= 7.854 \times 10^{-9} \text{ m}^4$$

Ball bearings would behave as simple supports for the shaft.
Length of the simply supported shaft (see Fig. 3.8), $l = 1$ m (given).

Finally, the critical speed of the system of Fig. 3.8 is given by Eq. (3.39):

$$N_c = \frac{60}{2\pi} \left(\frac{48EI}{ml^3} \right)^{1/2} = \frac{60}{2\pi} \left(\frac{48^*2^*10^{11*}7.854^*10^{-9}}{76.58^*(1)^3} \right)^{1/2}$$
$$= 299.6 \text{ RPM}$$

3.6 Dynamical Analogies

The electro-mechanical analogies and electro-acoustic analogies constitute a general class of dynamical analogies. These analogies are symbolic as well as physical (conceptual). In other words, these analogies imply similarity of the governing equations as well as physical behavior of the corresponding dynamical elements and state variables. Electromotive force (or voltage) and current in the electrical networks correspond to mechanical force and velocity in the mechanical vibrational systems. Similarly, electrical inductance, capacitance and resistance are analogous to inertia (or mass), spring compliance (inverse of stiffness) and damping, respectively [6]. Equations of dynamical equilibrium are found to be similar to loop equations in the analogous electrical networks, as illustrated in Fig. 3.9 for a two-DOF system.

The free-body diagrams of the lumped masses m_2 and m_4 yield the following equations of dynamical equilibrium:

$$m_2\ddot{x}_2 + c_1(\dot{x}_2 - \dot{x}_0) + k_1(x_2 - x_0) + c_3(\dot{x}_2 - \dot{x}_4) + k_3(x_2 - x_4) = 0$$
$$(3.40)$$

$$m_4\ddot{x}_4 + c_3(\dot{x}_4 - \dot{x}_2) + k_3(x_4 - x_2) = 0 \qquad (3.41)$$

For harmonic excitation (time dependence $e^{j\omega t}$), the time derivative d/dt is equivalent to a multiplication factor $j\omega$, and displacement

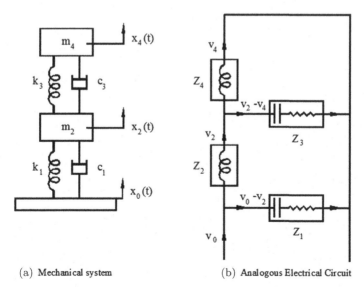

(a) Mechanical system   (b) Analogous Electrical Circuit

Fig. 3.9   An illustration of the electro-mechanical analogies: a system excited at the bottom support (constant velocity source or infinite-impedance source).

$x(t)$ and accelerations $\ddot{x}(t)$ are related to velocity $v = \dot{x}(t)$ as follows:

$$\ddot{x} = \frac{d^2x}{dt^2} = \frac{dv}{dt} = j\omega v \qquad (3.42)$$

Similarly,

$$x = \frac{v}{j\omega} \qquad (3.43)$$

Thus, the ordinary differential Eqs. (3.40) and (3.41) reduce to the following algebraic equations:

$$(j\omega m_2)v_2 + c_1(v_2 - v_0) + \frac{k_1}{j\omega}(v_2 - v_0) + c_3(v_2 - v_4)$$

$$+ \frac{k_3}{j\omega}(v_2 - v_4) = 0 \qquad (3.44)$$

$$(j\omega m_4)v_4 + c_3(v_4 - v_2) + \frac{k_3}{j\omega}(v_4 - v_2) = 0 \qquad (3.45)$$

These equations are analogous to Kirchhoff's loop equations for the electrical analogous circuit of Fig. 3.9(b):

$$Z_2 v_2 + Z_1(v_2 - v_o) + Z_3(v_2 - v_4) = 0 \qquad (3.46)$$

$$Z_4 v_4 + Z_3(v_4 - v_2) = 0 \qquad (3.47)$$

$$Z_2 = j\omega m_2, Z_4 = j\omega m_4, Z_1 = c_1 + \frac{k_1}{j\omega}, Z_3 = c_3 + \frac{k_3}{j\omega} \qquad (3.48)$$

Equations (3.48) indicate that mass is analogous to inductance, spring stiffness is analogous to reciprocal of capacitance, and damping coefficient is analogous to resistance, respectively, in the electrical networks.

It may be noted from Fig. 3.9 that the free end of mass m_4 in Fig. 3.9(a) is represented as a zero force or zero impedance or a short circuit in Fig. 3.9(b). As the force across a spring is proportional to differential displacement at its ends, it is represented as a shunt element. Similarly, as the force across a damper is proportional to differential velocity at its ends, it is also represented as a shunt element. However, the spring and damper that are parallel to each other and share the same terminations or ends, are in series with each other within the same shunt impedance in Fig. 3.9(b) (see Z_1 and Z_3).

Obviously, dynamical analogies illustrated in Fig. 3.9 provide an alternative analytical tool for evaluation of forced or steady-state response of multi- as well as single-DOF dynamical systems. They also help in deriving the natural frequency equation of a freely vibrating system.

An important advantage of dynamical analogies is that they help in better conceptualization or understanding of the physical function of different dynamical elements. This in turn helps in synthesis of vibration absorbers as well as isolators for different applications, as will become clear in the next chapter.

3.7 Vibration of Beams and Plates

Noise emitted by vibrating bodies (mostly thin plates and sheet-metal components) is one of the major sources of industrial as well as automotive noise. Control of this type of noise requires knowledge of the free as well as forced vibration of plates and plate-like surfaces. The equations governing the flexural vibration of plates may be conceptualized and developed as two-dimensional (2D) extension of the corresponding equations for transverse vibration of uniform beams where the effects of rotary inertia and shear deformation are neglected.

For a freely vibrating beam, the loading on the beam (see Fig. 3.10 for nomenclature and positive directions of the transverse displacement w, shear force V and bending moment M) will be equal to the inertia force. Thus,

$$EI\frac{\partial^4 w}{\partial x^4} = -\rho A\frac{\partial^2 w}{\partial t^2} \qquad (3.49)$$

where E is the Young's modulus of the beam material, I is the second moment of area of the cross-section about the neutral plane axis of the beam, ρ is the density and A is the cross-sectional area of the beam. Thus ρA represents the mass per unit length of the beam. The product EI is often referred to as flexural rigidity or flexural stiffness of the beam.

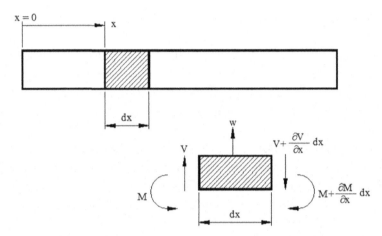

Fig. 3.10   Positive directions of the transverse displacement w, bending moment M and shear force V.

For harmonic time dependence,

$$w(x,t) = w(x)e^{j\omega t} \tag{3.50}$$

Equation (3.49) reduces to the ordinary differential equation (ODE):

$$EI\frac{d^4 w(x)}{dx^4} - \rho A\omega^2 w(x) = 0 \tag{3.51}$$

This linear, fourth-order ODE with constant coefficients has a general solution,

$$w(x) = C_1 e^{-jk_b x} + C_2 e^{jk_b x} + C_3 e^{-k_b x} + C_4 e^{k_b x} \tag{3.52}$$

where k_b, the bending wave number, is given by

$$k_b = \left(\frac{\rho A\omega^2}{EI}\right)^{1/4} \tag{3.53}$$

The four terms constituting Eq. (3.52) may be shown to represent the forward progressive wave, rearward progressive wave, the forward evanescent wave, and the rearward evanescent wave, respectively. The evanescent waves are produced at the excitation point, support, boundary and discontinuity, and decay exponentially without propagating. Therefore, the evanescent terms are not waves in a rigorous sense inasmuch as wave is defined as a moving disturbance. The evanescent terms are near-field effects.

Comparing the bending wave number k_b given by Eq. (3.53) with the acoustic wave number $k = \omega/c$ (see Eq. (1.6)), the bending wave speed is given by [1]:

$$c_b = \frac{\omega}{k_b} = \left(\frac{EI\omega^2}{\rho A}\right)^{1/4} \tag{3.54}$$

It may be noted that c_b is a function of frequency, unlike the sound speed c which is independent of frequency. Thus, different harmonics of a flexural disturbance will move at different speeds along the beam. This phenomenon is called Dispersion. In other words, unlike sound waves, the flexural or bending waves are dispersive.

Constants C_1–C_4 and wave number k_b (and thence the natural frequency ω) may be determined from the boundary conditions. For example, for the cantilever beam shown in Fig. 3.11, the boundary conditions are as follows:

At the fixed (clamped) end, $x = 0$:

$$\text{Displacement } w(0) = 0 \text{ and slope } \frac{dw}{dx}(0) = 0 \tag{3.55}$$

At the free end, $x = l$:

$$\text{Bending moment } EI\frac{d^2w(l)}{dx^2} = 0 \text{ and shear force } EI\frac{d^3w(l)}{dx^3} = 0 \tag{3.56}$$

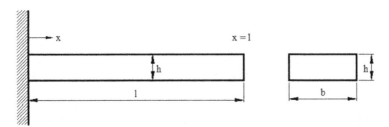

Fig. 3.11 Flexural vibration of a uniform cantilever beam of length l, width b and thickness h.

Thus, it can be shown that the frequency equation is a transcendental equation [1]:

$$\cos(k_b l)\cosh(k_b l) = -1 \qquad (3.57)$$

This has infinite number of roots, each of which corresponds to a natural frequency and a corresponding mode shape. Thus, displacement $w(x)e^{j\omega t}$ in response to an external transverse force $F(x)e^{j\omega t}$ may be expressed as a sum of all individual modes:

$$w(x) = \sum_{i=1}^{\infty} A_i \phi_i(x) \qquad (3.58)$$

where ϕ_i, the ith mode shape of the cantilever of Fig. 3.11, is given by [1, 7]:

$$\phi_i(x) = \cosh(k_{bi}x) - \cos(k_{bi}x) - \left\{\frac{\cosh(k_{bi}l) + \cos(k_{bi}l)}{\sin h(k_{bi}l) + \sin(k_{bi}l)}\right\}$$
$$\times \left\{\sinh(k_{bi}x) + \sin(k_{bi}x)\right\} \qquad (3.59)$$

Coefficient A_i, representing the relative strength or amplitude of the ith mode, may be found by making use of the orthogonality of natural modes.

In practice, the natural frequencies, modal shapes and response of a beam to a forcing function are found numerically on a digital computer, making use of the finite element model (FEM) and/or measurements.

Figure 3.12 shows the first three bending mode shapes for the cantilever of Fig. 3.11. The corresponding natural bending wave numbers are also indicated therein. The ith natural frequency can be found by means of Eq. (3.53). Thus,

$$f_{bi} = \frac{k_{bi}^2}{2\pi}\left(\frac{EI}{\rho A}\right)^{1/2} = \alpha_i \left(\frac{EI}{\rho A l^4}\right)^{1/2} \qquad (3.60)$$

where $\alpha_1 = 0.5595, \alpha_2 = 3.507$ and $\alpha_3 = 9.82$ for the first three natural modes of the cantilever vibration shown in Fig. 3.11 [1].

Relations (3.49) to (3.60) for flexural vibration of a beam find their counterparts in vibration of thin rectangular plates as follows.

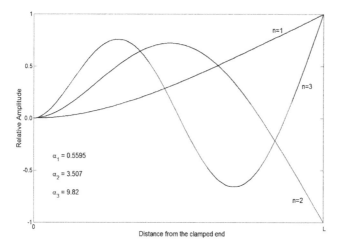

Fig. 3.12 The first three normal modes of vibration of the cantilever beam of Fig. 3.11.

The 2D wave equation for flexural or transverse vibration of a thin rectangular plate is

$$\frac{Eh^3}{12(1-v^2)}\left\{\frac{\partial^4 w}{\partial x^4} + 2\frac{\partial^4 w}{\partial x^2 \partial y^2} + \frac{\partial^4 w}{\partial y^4}\right\} + \rho h \frac{\partial^2 w}{\partial t^2} = 0 \qquad (3.61)$$

where h is the plate thickness (assumed to be uniform), v is Poisson's ratio, ρ is the density, so that the product ρh represents the mass per unit area (also called surface density) of the plate, and $w = w(x, y, t)$ is the transverse plate displacement.

Assuming time dependence to be $e^{j\omega t}$ as before, Eq. (3.61) reduces to a linear, fourth-order ODE, which may be solved by means of separation of variables. Then, making use of the appropriate boundary conditions for a rectangular plate of dimensions l_x and l_y, one gets the following relationships [1, 7]:

$$c_{Lp} = \left\{\frac{E}{\rho(1-v^2)}\right\}^{1/2} \qquad (3.62)$$

$$c_{bp} = \left\{\frac{Eh^3\omega^2}{12(1-v^2)\rho h}\right\}^{1/4} = (1.8c_{Lp}hf)^{1/2} \qquad (3.63)$$

$$k_{bp} = (k_x^2 + k_y^2)^{1/2} \qquad (3.64)$$

Here, c_{Lp} is the quasi-longitudinal wave velocity, c_{bp} is the bending wave velocity for thin plate, k_x and k_y are the x-component and y-component of k_{bp}, the bending wave number (considered a vector).

For a simply supported plate,

$$k_x = \frac{m\pi}{l_x} \quad \text{for } m = 1, 2, 3, \ldots \tag{3.65}$$

$$k_y = \frac{n\pi}{l_y} \quad \text{for } n = 1, 2, 3, \ldots \tag{3.66}$$

$$f_{m,n} = (1.8 c_{Lp} h) \left\{ \left(\frac{m}{2l_x} \right)^2 + \left(\frac{n}{2l_y} \right)^2 \right\}, \text{ Hz} \tag{3.67}$$

where m and n are the number of half-waves in the x- and y-directions, and $f_{m,n}$ is the natural frequency corresponding to the (m, n) mode of vibration.

The corresponding relationships for a rectangular plate clamped on all four edges are [1, 7]:

$$k_x = \frac{(2m + 1)\pi}{l_x} \quad \text{for } m = 1, 2, 3 \ldots \tag{3.68}$$

$$k_y = \frac{(2n + 1)\pi}{l_y} \quad \text{for } n = 1, 2, 3 \ldots \tag{3.69}$$

$$f_{m,n} = (1.8 c_{Lp} h) \left\{ \left(\frac{2m + 1}{2l_x} \right)^2 + \left(\frac{2n + 1}{2l_y} \right)^2 \right\}, \text{ Hz} \tag{3.70}$$

Equations (3.62)–(3.70) indicate that the quasi-longitudinal wave speed along a plate is slightly higher than the longitudinal wave speed along a slender rod. Besides, natural frequency $f_{m,n}$ of a plate clamped along all four edges is considerably higher than that of a plate simply supported along all the edges.

In practice, vibrating surface of a machine cannot be idealized as rectangular thin plate with boundary conditions of clamping or simple support. Then, one resorts to numerical analysis on a digital computer making use of FEM and/or measurements, where use is made of the orthogonality of normal modes. This is particularly true when one tries to evaluate sound power radiated by the surfaces of a typical machine-like engine, gear box, automobile, etc., excited by multifarious time-variant forces.

3.8 Vibration Measurement

For harmonic time dependence ($e^{j\omega t}$), displacement ξ, velocity u and acceleration a are related as follows:

$$u = j\omega\xi \quad \text{and} \quad a = j\omega u = -\omega^2\xi \tag{3.71}$$

Thus, amplitudes of u and a would be ω and ω^2 times that of the corresponding displacement. For a constant velocity spectrum, the displacement would decrease linearly with frequency and acceleration would increase linearly with frequency, as shown in Fig. 3.13.

Therefore, in general, displacement transducer is used for low-frequency measurements and acceleration transducer (accelerometer) is preferred for high-frequency measurements. For mid-frequency measurements and in general for wide-frequency spectrum measurements, velocity pickup may be used. However, accelerometer is the most commonly used vibration transducer; it has the best all-round characteristics.

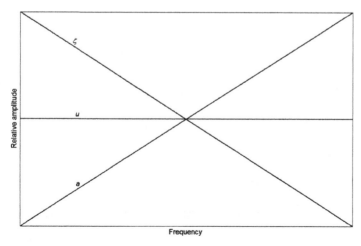

Fig. 3.13 Relative amplitudes of displacement (ξ), velocity (u) and acceleration (a) (for constant velocity).

Often, vibration, like sound, is measured in logarithmic units, decibels (dB), as follows:

$$\text{Displacement level, } L_d = 20 \log \left(\frac{\xi_{rms}}{\xi_{ref}} \right), \text{ dB} \qquad (3.72)$$

$$\text{Velocity level, } L_u = 20 \log \left(\frac{u_{rms}}{u_{ref}} \right), \text{ dB} \qquad (3.73)$$

$$\text{Acceleration level, } L_a = 20 \log \left(\frac{a_{rms}}{a_{ref}} \right), \text{ dB} \qquad (3.74)$$

Here, as per international standards, the reference values are [1]:

$$\xi_{ref} = 10^{-12} \text{m} = 1 \,\text{pm} \qquad (3.75)$$

$$u_{ref} = 10^{-9} \text{m/s} = 1 \,\text{nm/s} \qquad (3.76)$$

$$a_{ref} = 10^{-6} \text{m/s}^2 = 1 \,\text{micrometer/s}^2 \qquad (3.77)$$

It may be noted that at $\omega = 1000 \,\text{rad/s}$, or at $f = 1000/2\pi = 159.155 \,\text{Hz}$, all three levels will be equal. This determines the frequency at which all three lines intersect in Fig. 3.13.

The reference values indicated in Eqs. (3.75)–(3.77) are not always adhered to in the existing literature. Therefore, it is advisable to indicate the reference value used in prescribing vibration levels; for example, velocity level $=75 \,\text{dB re } 10^{-9} \,\text{m/s}$.

Vibration levels may be added or subtracted in the same way as sound pressure levels, as illustrated before in Chapter 1.

Example 3.5. Vibration velocity levels were measured to be 100, 90, 80 and 70 dB in the Octave bands centered at 31.5, 63, 125 and 250 Hz, respectively. Evaluate the total rms values of displacement, velocity and acceleration.

Solution. We make use of Eqs. (3.72), (3.73) and (3.74) that relate the rms values of displacement, velocity and acceleration to the respective levels in dB, and Eq. (3.71) to evaluate displacement and acceleration from velocity.

Thus,

$$u_{rms} = 10^{-9} \cdot 10^{L_u/20}, \quad \xi_{rms} = \frac{u_{rms}}{\omega}, \quad a_{rms} = \omega \cdot u_{rms}, \quad \omega = 2\pi f$$

Octave band center frequency f (Hz)	31.5	63	125	250	Total
Velocity level L_u (dB)	100	90	80	70	100.5
u_{rms} (m/s)	1.0×10^{-4}	3.16×10^{-5}	1.0×10^{-5}	3.16×10^{-6}	1.06×10^{-4}
$\omega = 2\pi f$(rad/s)	196	393	785	1571	—
$\xi_{rms} = u_{rms}/\omega$ (m)	5.1×10^{-7}	8.04×10^{-8}	1.27×10^{-8}	3.01×10^{-9}	5.16×10^{-7}
$a_{rms} = \omega * u_{rms}$ (m/s^2)	0.0196	0.0124	0.0078	0.0050	0.025

In the foregoing table, the total velocity level is determined by means of Eq. (1.37), and the total root mean square value of velocity is evaluated as follows:

$$u_{rms}(total) = \left[\sum_{i=1}^{4} u_{rms}^2(f_i) \right]^{1/2}$$

Similar expressions have been used for ξ_{rms} (total) and a_{rms} (total).

Incidentally, it may be noted that total values in the last column are practically equal to those in the first column where the velocity level is 100 dB, i.e., higher than those in other octave bands by 10 dB or more. This observation may be used effectively for noise and vibration control as will be demonstrated in the subsequent chapters.

Depending on the mode of transduction, different types of vibration transducers are available in the market; namely, eddy current displacement probes, moving element velocity pickups, piezoelectric accelerometers, etc.

Eddy current displacement probes are non-contacting displacement transducers with no moving parts and work right down to zero frequency. However, their higher frequency limit is about 400 Hz because, as indicated above, displacement often decreases with frequency. Being of the non-contacting type, the eddy-current probes are ideally suited for rotating machinery. However, their dynamic range is limited to 100:1 or 40 dB ($20\log(100/1) = 40$ dB).

This limitation of dynamic range is also typical of the moving element velocity pick-ups. As these have to be in contact with the vibrating surface, their mass may alter the vibration that is supposed to be measured, particularly for thin sheet metal components. The lower frequency limit of these pick-ups is about 10 Hz because they operate above their mounted resonance frequencies. The moving elements are prone to wear and therefore the durability of these pickups is rather limited. Sensitivity to orientation and magnetic fields are additional concerns with the moving element velocity pick-ups.

Accelerometers have a very wide dynamic range (>120 dB) as well as frequency range, although they fail as frequency tends to zero. Accelerometers are rugged transducers. Mostly they are of the piezoelectric type where sensing element is a polarized piezoelectric crystal or ferroelectric ceramic element. An electric charge is produced when this element is stressed in shear or in tension/compression. Figure 3.14 shows a schematic of a compression-type piezoelectric accelerometer.

Acceleration of the mass element in Fig. 3.14 provides the inertial compressive force on the piezoelectric crystal, with the preloading spring ensuring that this force remains compressive throughout the oscillation. The compression-type accelerometer is generally used for measuring high shock levels, whereas the shear-type accelerometer serves a general purpose. Like velocity pick-up, accelerometer has

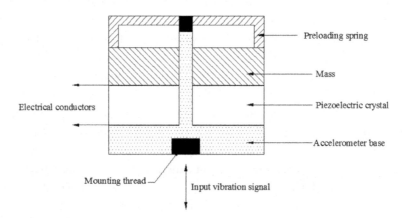

Fig. 3.14 Schematic of a compression-type piezoelectric accelerometer.

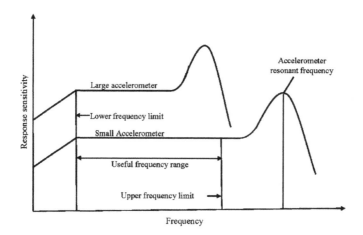

Fig. 3.15   Relative frequency response of accelerometers.

to be in contact with the measuring surface and therefore has to be designed to be light and small enough not to interfere with vibration that it is supposed to measure.

Accelerometer's useful frequency range is limited on the lower frequency side by decreased sensitivity and on the higher frequency side by the resonance frequency of the accelerometer, as shown in Fig. 3.15. This resonance frequency is the result of the preloading spring interacting with the inertial mass. In general, larger accelerometers has larger sensitivity but have a restricted useful frequency range. They may also interfere with the vibration surface through inertial loading as indicated above. Therefore, one must select a small and light accelerometer and then increase its electrical output through appropriate amplification [1].

3.9 Laser Doppler Vibrometer

A laser Doppler vibrometer (LDV) measures surface vibration velocity in a non-contact mode. The measured vibration data can be converted to displacement and acceleration through integration and differentiation. A Helium-Neon (He-Ne) laser is typically used for the LDV, which produces a visible red beam with a wavelength of $0.6328 \, \mu m$.

A laser vibrometer is generally a two-beam laser interferometer that measures the frequency or phase difference between an internal reference beam and a test beam. The test beam is directed towards the target, and scattered light from the target is collected and interfered with the reference beam on a photodetector. This vibrometer operates in a heterodyne regime by introducing a known frequency shift to one of the beams. This frequency shift is typically generated by a Bragg cell.

A schematic of how a laser vibrometer works is shown in Fig. 3.16. The laser beam, with a frequency f_r, is divided into a reference beam and a test beam using a beam-splitter. The test beam then passes through the Bragg cell, which adds a frequency shift f_b. This frequency-shifted beam is then directed towards the target. The motion of the target adds a Doppler shift f_d to the beam. Light scatters from the target in all directions, but a portion of the light is collected by the LDV, reflected by the beam-splitter, and detected by the photodetector. This scattered light has a frequency equal to $f_r + f_b + f_d$. The photodetector combines this scattered light with the reference beam [8]. The reflected beam frequency is shifted in proportion to the velocity of the object. Monitoring shifted frequency as a function of time provides the structure velocity waveform. The major advantage of the LDV is non-contact, and there is no mass loading

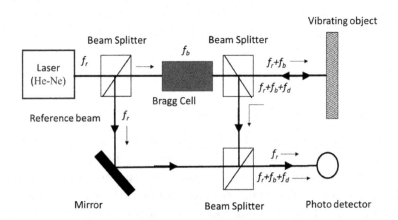

Fig. 3.16   Schematic diagram of LDV.

effect while doing the vibration measurement. The vibration measurements on the structure at multiple locations can be automated by "scanning laser vibrometer."

3.10 Measurement of Damping

Figure 3.1 shows damper in the form of a dashpot. However, in practice, damping is rarely introduced into a system in the form of a dashpot. Often, one makes use of viscoelastic layers like rubber mats and wedges, fluid-dynamic devices like the so-called shock absorbers, etc. Damping coefficient of such devices or the in-built structural damping is rarely known or given; it needs to be measured [5]. Often it is inferred indirectly from different manifestations of damping; viz.

(a) decreasing amplitude of the system in free vibration (see Fig. 3.2),

(b) continuous conversion of mechanical energy (potential as well as kinetic) into heat in the form of hysteresis losses, and

(c) resonance characteristics (the extent to which amplitude of vibration is limited at resonance and the half-power bandwidth of the resonance curve) during forced (steady state) vibration (see Fig. 3.3).

3.10.1 *Logarithmic decrement method*

Referring to the damped single-DOF system of Fig. 3.1 and the free vibration response shown in Fig. 3.2, the amplitude envelope is given by $x_0 e^{-\xi \omega_n t}$ when the mass is displaced from its mean position by x_0 and released without imparting any initial velocity to it. The ratio of amplitudes for two successive peaks would then be

$$\frac{X_i}{X_{i+1}} = \frac{x_0 e^{-\zeta \omega_n t}}{x_0 e^{-\zeta \omega_n (t+T)}} = e^{\zeta \omega_n T} = e^{\zeta \omega_n 2\pi / \omega_d}$$

$$= e^{2\pi \zeta / (1-\zeta^2)^{1/2}} \simeq e^{2\pi \zeta} \tag{3.78}$$

because the damping ratio ζ is often much less than unity.

A popular measure of damping is Logarithmic Decrement, δ, defined as natural logarithm of the ratio of successive amplitudes.

Thus,

$$\delta = \ln \left(\frac{X_i}{X_{i+1}} \right) \simeq 2\pi\zeta \qquad (3.79)$$

Obviously, δ is easy to measure. Then, Eq. (3.79) can be used to get damping ratio, ζ. Finally, damping coefficient c may be determined from Eq. (3.2), i.e.

$$c = c_c\zeta = 2m\omega_n\zeta \qquad (3.80)$$

3.10.2 *Half-power bandwidth method*

It may be noted from the steady-state response curves of Fig. 3.3 that with increasing damping, the resonance amplitude decreases, and the curve becomes more flat. This flat-ness, in terms of half-power bandwidth, may be used to evaluate the damping ratio ζ and thence the damping coefficient c. In fact, this method is a standard method for evaluation of the inherent or structural damping of materials (beams and plates) in flexural mode of vibration in terms of the loss factor η, which is a measure of to what extent strain fluctuations would lag behind the corresponding stress fluctuations. It is defined in terms of the complex character of the Young's modulus:

$$E = E_r + jE_i = E_r(1 + j\eta), \eta \equiv E_i/E_r \qquad (3.81)$$

where E_r is called the Storage Modulus, E_i is called the Loss Modulus, and η is called the Loss Factor.

For rubber-like viscoelastics E_r, E_i and η are functions of frequency. These are evaluated by the standard Oberst's beam method [9].

When the system (or a cantilever beam) is excited by means of a wide-band random excitation, the typical response (vibration amplitude) consists of several peaks, one of which is shown in Fig. 3.17.

If f_0 is a resonance frequency with peak amplitude of X_0 and f_1 and f_2 are the frequencies on either side of f_0, as shown in Fig. 3.17, where $X_1 = X_2 = 0.707X_0$ (the power is proportional to vibration

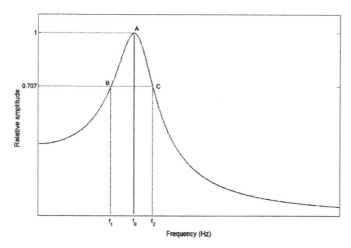

Fig. 3.17   Illustration of half-power bandwidth (points B and C are half-power points).

amplitude squared), then,

$$\text{Normalized half-power bandwidth, } bw \equiv \frac{f_2 - f_1}{f_0} \qquad (3.82)$$

It may be shown that for small values of damping ($\zeta^2 \ll 1$), different measures of damping are interrelated as follows [10]:

$$\zeta = \frac{\eta}{2} = \frac{bw}{2} = \frac{\delta}{2\pi} = \frac{1}{2Q} \qquad (3.83)$$

Here Q is called the quality factor (in Electrical Filter theory), representing the non-dimensional amplitude or dynamic magnification factor $X/(F_0/k)$ at resonance $\omega = \omega_0$ (or $f = f_0$).

Example 3.6. In an Oberst Beam test, the normalized half-power bandwidth of the first resonance peak is measured to be 0.16 for a particular alloy steel. If the same cantilever beam were disturbed and left to vibrate freely, what would be the ratio of two successive amplitudes of the beam in free vibration?

Solution. The normalized half-power bandwidth bw is given to be 0.16.

Making use of Eq. (3.83), the logarithmic decrement

$$\delta = \pi * bw = \pi * 0.16 = 0.5$$

Now, by definition of logarithmic decrement, the ratio of two successive amplitudes, X_i/X_{i+1} is given by Eq. (3.79):

$$\frac{X_i}{X_{i+1}} = e^\delta = e^{0.5} = 1.65$$

Incidentally, making use of Eq. (3.83) again, the loss factor η of the alloy steel is given by

$$\eta = bw = 0.16$$

By comparison, loss factor of normal steel is of the order of 0.001. Thus, the loss factor of this alloy steel is 160 times that of ordinary steel. Therefore, the resonance amplitude of a plate made out of this alloy steel will be only 1/160 times that of normal steel plate. In other words, the alloy steel plate will be quieter by

$$20 \log 160 = 44.1 \, \mathrm{dB}$$

in free or resonant vibration, provided it has the same storage modulus E_r.

This is why sheet metal components of engines, compressors, etc. should be made out of materials with good inherent structural damping (high loss factor η), or better still, high loss modulus $E_i = \eta E_r$ (see Eq. 3.81)).

References

[1] Norton, M. P., *Fundamentals of Noise and Vibration*, Cambridge University Press, Cambridge (1989).
[2] Yang, B., *Theory of Vibration – Fundamentals, in Encyclopedia of Vibration* (Ed. S. G. Braun), Academic Press, San Diego (2002), pp. 1290–1299.
[3] Lieven, N. A. J., Forced Response, *Encyclopedia of Vibration* (Ed. S. G. Braun), Academic Press, San Diego (2002), pp. 578–582.
[4] Irwin, J. D. and Graf, E. R., *Industrial Noise and Vibration Control*, Prentice Hall, Englewood Cliffs (1979).

[5] Blevins, R. D., *Formulas for Natural Frequency and Mode Shape*, Van Nostrand Reinhold Co., New York (1979).

[6] Olson, M. F., *Dynamical Analogies*, Second Edition, Van Nostrand, Princeton (1958).

[7] Norton, M. P. and Drew, S. J., Radiation by flexural elements, in *Encyclopedia of Vibration* (Ed. S. G. Braun), Academic Press, San Diego (2002), pp. 1456–1480.

[8] Laser vibrometry: An introduction to non-contact vibration measurement. http://Polytec.com.

[9] Anon, Standard test method for measuring vibration-damping properties of materials, ASTM International Standard, E 756-04 (2004).

[10] Ewins, D. G., Damping measurement, in *Encyclopedia of Vibration* (Ed. S. G. Braun), Academic Press, San Diego (2002), pp. 332–335.

Problems

3.1. If the machine of the single-DOF system of Example 3.2 ($m = 500\,\text{kg}$, $k = 50968\,\text{N/m}$) were excited by a vertical unbalanced force of 100 N at a rotational speed of 1500 RPM (revolution for minute), then

(a) what would be the amplitude of vibration of the machine, and

(b) what would be amplitude of the force transmitted to the (rigid) foundation?

[Ans.: (a) 8.14 microns, (b) 0.4153 N]

3.2. A single–DOF spring-mass-damper system shown in Fig. 3.1 has the following parameters: mass $m = 6\,\text{kg}$, spring constant, $k = 15,000\,\text{N/m}$. Determine the damping coefficient, c of the dashpot from the observation that in free vibration, amplitude of oscillation of the mass decreases to 20% of its displacement in seven consecutive cycles.

[Ans.: 21.96 Ns/m]

3.3. If the system of Fig. 3.1 is executing free vibrations and its displacement $x(t)$ is given by the expression,

$$x(t) = 0.002e^{-6t}(\sin 10.4t + 1.732\cos 10.4t),$$

determine the undamped natural frequency ω_n and damping ratio ζ of the system.

[Ans.: 13.0 rad/s and 0.5]

3.4. For the system of Fig. 3.1, let $m = 100\,\text{kg}$, $k = 100\,\text{N/mm}$, $c = 10\,\text{Ns/cm}$. Evaluate its damping ratio, undamped natural frequency, and amplitude at resonance for the force amplitude of 100 N.

[Ans.: 0.158, 31.6 rad/s and 3.16 mm]

3.5. The root-mean-square vibration acceleration values measured in different octave bands are listed below:

Octave band center frequency (Hz)	31.5	63	125	250	500	1000
RMS vibration acceleration (m/s^2)	5.0	4.0	3.0	3.0	1.0	0.5

Calculate

(a) overall RMS acceleration in m/s^2 and decibels.
(b) overall RMS velocity in mm/s and decibels.
(c) overall RMS displacement in μm and decibels.

[Ans.: (a) 7.433 m/s^2 and 137.4 dB; (b) 27.54 mm/s and 148.8 dB; (c) 130.4 μm and 162 dB]

Chapter 4

Vibration Control

4.1 Introduction

A vibration problem generally involves a source of vibration, a dynamical (or vibratory) system and the response. The system may be looked at as "transmission path" as is done in the noise control practice (see Chapter 8). Control of vibration, therefore, consists in modifying the source and/or the system.

Typical sources of vibration are unbalanced forces and moments in a reciprocating and rotating machinery, turbulent flow in pipes (particularly, at the bends and intersections), standing waves and surges. Therefore, vibration control at the source involves balancing of reciprocating and rotating machinery (engines, turbines, compressors, blowers, etc.), smoothening of flow in pipes (avoiding separation of boundary layer), proper lubrication at joints, etc. Often, there is a coupling between vibration and noise; reducing one would result in reduction of the other. That is why the vibration control and noise control are considered together in this textbook.

The system consists of inertias, isolator springs and dampers. Control of vibration involves proper design of the system (or transmission path) so as to reduce the vibration of machinery as well as transmissibility over the entire range of speeds or forcing frequencies.

Permissible vibration level for each part of a system in a dynamic setting is decided by requirements of functionality or comfort. International Standards Organization (ISO) provides various regulations and standards which are usually stated in terms of amplitude, frequency and, sometimes, duration of test. Figure 4.1 is representative of the acceptable vibration levels.

As indicated in Chapter 3, vibration control requirement occurs in two classes of vibration: (a) isolation of a sensitive instrument from the support or base motion, and (b) isolation of a support or foundation from the unbalanced forces generated within a machine. Base excitation occurs during vehicle motion over an undulating surface, satellite launch, and in the operation of disk drives. The force transmissibility problem relates to machine mounts, engine mounts, machine tool vibration, etc.

It is generally cost-effective to control vibration at the source. Theory of vibration outlined in Chapter 3 suggests a number of

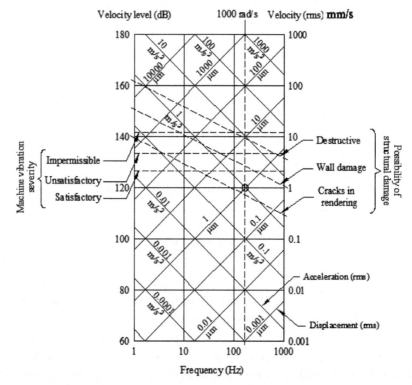

Fig. 4.1 Acceptable vibration levels. (Note that at 159.15 Hz or 1000 rad/s, displacement level, velocity level and acceleration level will all be equal to 120 dB, as indicated earlier in Fig. 3.13.)

vibration control measures. These are discussed at some length in the following sections.

4.2 Vibration Control at the Source

As indicated above, vibration can be controlled at the source by reducing the excitation. This may be done by

(a) reducing the rotational speed of the machine, if possible, without compromising the primary function of the machine,
(b) reducing the unbalance, which in turn may be affected by

    (i) precise machining of the rotor or crankshaft, and
    (ii) selecting a proper configuration (e.g., 6-cylinder inline engine is inherently balanced).

Similarly, the flow excitation may be reduced by

(a) reducing the flow speed by reducing the flow and/or increasing the area of cross-section,
(b) smoothening the flow by means of tubular flow straighteners,
(c) streamlining the flow or avoiding the boundary layer separation by means of guide vanes.

Often, tall vertical chimneys or stacks are set into violent vibration by transverse oscillating forces caused by vortex shedding (Karman vortices) during strong wind. This may be effectively avoided by means of helical spoilers around the chimney that break the regular vortex pattern in the wake, thereby reducing the excitation dramatically. This is a good example of vibration control at the source. A variation of the same principle is made use of in order to minimize sloshing in large vertical, cylindrical liquid reservoirs. Perforated vertical plates are placed in the reservoir, parallel to, and some distance away from the container wall.

One common source of vibration (and noise) is dry friction caused by failure of lubrication. This causes high-frequency self-excitation. This may be avoided or remedied by providing proper grease cups and keeping them under regular inspection.

A rotating shaft with a keyway causes parametric excitation due to periodical variation in transverse flexural stiffness. This may be remedied by providing two identical keyways on either side of the original keyway at 120° azimuthal locations. A precision job would completely nullify the parametric excitation that could have caused a whirling motion of the rotor, as explained in Chapter 3.

In general, the unbalanced forces and moments of a reciprocating engine crankshaft and turbine or compressor rotor cannot be reduced to zero. Therefore, balancing has to be done at site and/or at the manufacturer's end. Several techniques and machines have been developed over the years for balancing of rigid rotors. These include [1]:

(a) pivoted carriage balancing machine,
(b) Gisholt-type balancing machine.

A very large rotor cannot be mounted on any balancing machine. However, portable sets for carrying out the field balancing of both single and two-plane rotors are commercially available.

A rotor operating at a speed higher than its first critical speed undergoes a significant transverse deflection at this speed, and therefore is termed as flexible rotor. In a rigid rotor, the balancing masses are attached to neutralize the unbalanced forces and moments. However, in a flexible rotor, the balancing masses are designed to suitably modify the dynamic deflection characteristics of the rotor. This is affected by means of the modal balancing technique [2].

4.3 Vibration Isolators

Theory of vibration of single-degree-of-freedom systems has been discussed in Chapter 3. Salient conclusions of the same regarding vibration isolation were as follows:

(i) For an isolator to yield low transmissibility, the natural frequency of the system should be much less than the forcing frequency, which is often equal to the rotational speed of the machine in corresponding units.

(ii) A sufficiently low natural frequency can be achieved by means of soft springs and/or inertia block.

(iii) Use of inertia block has the additional advantage of limiting the vibration levels of the machine.

(iv) Vibration of the machine and transmissibility at and around the resonance frequency may be controlled by means of damping. Excessive damping could, however, increase transmissibility at higher frequencies. Viscoelastic elements provide compliance as well as damping.

4.3.1 *Bonded rubber springs*

Natural frequency is inversely proportional to the square root of static deflection. The requirement of sufficiently low natural frequency results in large static deflection. Figure 4.2 indicates the range of static deflection for different types of isolators.

Large static deflection results in mechanical instability. This may be avoided by means of bonded rubber springs. Most anti-vibration

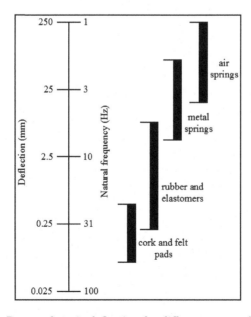

Fig. 4.2   Range of static deflection for different types of isolators.

mounts (AVMs) in the market make use of these springs. As its name suggests, a bonded rubber spring is constructed by bonding the rubber to metal parts. In fact, one can make use of several of them in series, as shown schematically in Fig. 4.4. It may be noted that four identical rubber pads have been bonded to five thin metal plates. Stiffness of such a composite spring in compression will be roughly one-fourth of that of a single rubber pad, EA/h, where A is area of cross-section of the spring. The same would apply to the spring stiffness in shear. Thus, for the composite rubber spring of Fig. 4.3,

Compressive or axial stiffness,

$$k_a \approx \frac{EA}{4h} \tag{4.1}$$

Torsional or shear stiffness,

$$k_s \approx \frac{GA}{4h} \tag{4.2}$$

Here, h is thickness of each of the four rubber pads. It may be noted that thickness of thin metal plates does not influence the axial as

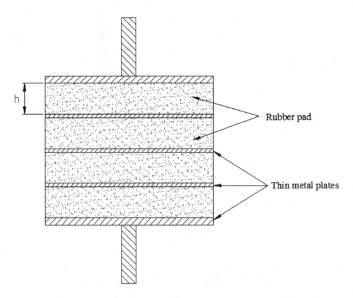

Fig. 4.3   A composite bonded rubber spring.

well as shear stiffness of the composite bonded rubber spring shown in Fig. 4.3 because stiffness of the metallic sheet is higher than that of the rubber pad by several orders.

Example 4.1. A studio of total mass 40.5 tons is to be supported on nine composite bonded rubber springs of the type shown in Fig. 4.3 in order to ensure a floating floor with natural frequency of 8 Hz. This is required to avoid flanking transmission of structure-borne sound at 32 Hz upwards. If the surface area of each of the four rubber pads in each of the nine springs is 0.3 m × 0.3 m and elastic modulus of rubber is 10^8 N/m^2, then evaluate thickness of each of the rubber pads. Assuming loss factor of rubber to be 0.2, evaluate transmissibility.

Solution. Stiffness of each spring

$$= \frac{EA}{4h} = \frac{10^8 * (0.3 * 0.3)}{4h} = \frac{0.0225 * 10^8}{h} \text{N/m}^2$$

Total stiffness of all nine springs,

$$k = \frac{9 * 0.0225 * 10^8}{h} = \frac{0.2025 * 10^8}{h} \text{N/m}^2$$

Total mass of the studio, $m = 40.5\,T = 4.05 \times 10^4$ kg
Natural frequency of the floating studio,

$$f_n = \frac{1}{2\pi} \left(\frac{k}{m} \right)^{1/2}$$

Thus,

$$\frac{1}{2\pi} \left(\frac{0.2025 * 10^8}{h * 4.05 * 10^4} \right)^{1/2} = 8$$

which gives pad thickness,

$$h = \frac{0.2025 * 10^8}{64 * 4\pi^2 * 4.05 * 10^4} = 0.198 \text{ m} = 198 \text{ mm}$$

This is unusually large from the mechanical stability point of view. It would be better to replace four pads of 198 mm thickness with eight pads of 99 mm thickness each, so that total static deflection remains unchanged.

As per Eq. (3.83),

$$\text{damping ratio}, \zeta = \frac{\eta}{2} = \frac{0.2}{2} = 0.1$$

$$\text{frequency ratio}, r = \frac{f}{f_n} = \frac{32}{8} = 4$$

Finally, use of Eq. (3.33) gives,

$$\text{Transmissibility } TR = \left[\frac{1 + (2 \times 0.1 \times 4)^2}{(1 - 4^2)^2 + (2 \times 0.1 \times 4)^2} \right]^{1/2} = 0.085 = 8.5\%$$

Incidentally, without damping, the transmissibility would be

$$TR(\text{without damping}) = \left| \frac{1}{1 - 16} \right| = 0.0667 = 6.67\%$$

Bonded rubber springs are available in various forms. Each is configured for a particular type of loading. For combined compression and shear loading, one uses Anti Vibration Mount (AVM) shown in Fig. 4.4.

Its stiffness for vertical loading is given by

$$k = 2A(G \sin^2 \alpha + E_a \cos^2 \alpha)/h \tag{4.3}$$

Here, A is the cross-sectional area of each rubber pad normal to the dimension h, and E_a is the apparent Young's modulus in compression,

$$E_a = 6.12G \tag{4.4}$$

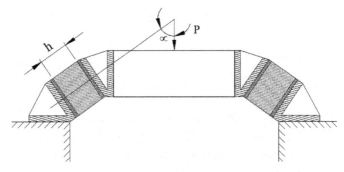

Fig. 4.4   AVM for combined compression and shear loading [1].

Poisson's ratio, v, for rubber is nearly 0.5. Therefore, Young's modulus for rubber is given by

$$E = 2(1 + v)G = 3G \tag{4.5}$$

Thus, the constraining effect of the end metal plates is to increase the apparent Young's modulus E_a to more than double its basic value. The angle α in Fig. 4.4 is determined from the consideration of limiting the compressive strain to 0.2 and the shear strain to 0.36. If x is the vertical deflection of the spring, then the accompanying shear deflection $x_s = x \sin \alpha$, and the compressive deflection $x_s = x \cos \alpha$. Therefore, the shear strain is $x \sin \alpha / h$ and the compressive strain is $x \cos \alpha / h$. In order to ensure that the two types of strain reach their limits simultaneously, α is fixed at arctan $(0.36/0.2)$ or $60°$ [1].

The AVMs of the type of Fig. 4.4 are used extensively under stationary installations like diesel generator (DG) sets, turbogenerators (TGs), stationary compressors, etc. Often, several AVMs are used in parallel such that their centre of buoyancy coincides with the centre of gravity of the machine.

Example 4.2. Evaluate the stiffness of the AVM of Fig. 4.4 for $\alpha \approx 60^0$, $h = 5$ cm, $A = 100$ cm^2 and $G = 10^6$ Pa.

Solution. Making use of Eqs. (4.4) and (4.3) we get,
Apparent Young's modulus, $E_a = 6.12 \times 10^6$ Pa
Stiffness, $k = 2 * \frac{100}{100*100}(10^6 \sin^2(\pi/3) + 6.12 * 10^6 \cos^2(\pi/3))/\frac{5}{100}$

$$= 2 \times 10^6 \left(\frac{3}{4} + 6.12 \left(\frac{1}{4} \right) \right) \Big/ 5$$

$$= 0.912 \times 10^6 \text{ Pa.m}$$

$$= 912 \text{ kN/m}$$

$$= 912 \text{ N/mm}$$

4.3.2 *Effect of compliant foundation*

In the theory of vibration dealt with in Chapter 3, the support or foundation was implicitly assumed to be rigid. However, in practice, the foundation or support would invariably have a finite impedance. In other words, it would be compliant, not rigid.

Making use of the electro-mechanical analogies [3] discussed in Chapter 3, for the dynamical system of Fig. 4.5 we have,

$$Z_1 = j\omega m, \quad Z_2 = \frac{k}{j\omega}, \quad (v_1 = \dot{x}_1, v_f = \dot{x}_f) \tag{4.6}$$

$$Z_1 v_1 + Z_2(v_1 - v_f) = F \tag{4.7}$$

$$Z_f v_f - Z_2(v_1 - v_f) = 0 \tag{4.8}$$

Simultaneous solution of the linear algebraic equations (4.7) and (4.8) yields,

$$v_f = \frac{Z_2 F}{Z_f(Z_1 + Z_2) + Z_1 Z_2} \tag{4.9}$$

Therefore, the force transmitted to the foundation and transmissibility are given by

$$F_f = Z_f v_f = \frac{Z_f Z_2 F}{Z_f(Z_1 + Z_2) + Z_1 Z_2} \tag{4.10}$$

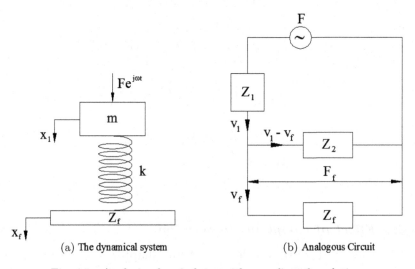

(a) The dynamical system   (b) Analogous Circuit

Fig. 4.5   Analysis of an isolator with compliant foundation.

$$TR = \left| \frac{F_f}{F} \right| = \left| \frac{Z_f Z_2}{Z_f(Z_1 + Z_2) + Z_1 Z_2} \right|, \ Z_1 = j\omega m, \ Z_2 = \frac{k}{j\omega}$$

$$(4.11)$$

As a consistency check, for a rigid foundation ($Z_f \to \infty$), Eq. (4.11) reduces to the following:

$$TR = \left| \frac{Z_2}{Z_1 + Z_2} \right| = \left| \frac{k/j\omega}{j\omega m + k/j\omega} \right| = \left| \frac{1}{1 - m\omega^2/k} \right| = \left| \frac{1}{1 - (\omega/\omega_n)^2} \right|$$

$$(4.12)$$

If the foundation could be idealized as a free mass M, then $Z_f = j\omega M$, and Eq. (4.11) would yield,

$$TR = \left| \frac{(j\omega M(k/j\omega))}{j\omega M(j\omega m + k/j\omega) + (j\omega m)(k/j\omega)} \right| = \left| \frac{1}{1 - m\omega^2/k + m/M} \right|$$

$$(4.13)$$

TR would tend to infinity at the natural frequency of the system, given by

$$1 - \frac{m\omega_n^2}{k} + \frac{m}{M} = 0$$

whence

$$\omega_n = \left(\frac{k(M+m)}{Mm} \right)^{1/2} = \left\{ \frac{k}{m} \left(1 + \frac{m}{M} \right) \right\}^{1/2} \qquad (4.14)$$

Thus, the natural frequency of a system with free inertial foundation would be higher than that with a rigid foundation.

For a compliant foundation, the real effectiveness of a vibration isolator (spring of stiffness k in Fig. 4.5) is not represented by transmissibility. A more appropriate metric in this case would be the ratio of the force transmitted to the foundation with the spring and that without the spring ($Z_2 \to \infty$). Let us denote it by force ratio, FR. Thus,

$$FR = \left| \frac{F_f \text{ with isolator spring}}{F_f \text{ without isolator spring}} \right| = \left| \frac{\frac{Z_f Z_2 F}{Z_f(Z_1+Z_2)+Z_1 Z_2}}{\frac{Z_f F}{Z_1+Z_f}} \right| \qquad (4.15)$$

This simplifies to

$$FR = \left| \frac{Z_1 + Z_f}{Z_1 + Z_f(1 + Z_1/Z_2)} \right| \tag{4.16}$$

It may be noted that for a rigid foundation ($Z_f \to \infty$), force ratio FR reduces to Transmissibility TR, i.e., Eq. (4.12). In other words, only for rigid support or foundation, do the two metrics match.

Example 4.3. The floating floor construction of Example 4.1 presumes a rigid foundation. If the foundation were compliant and could be modeled as a free mass of 200 tons, then what would be the transmissibility and the force ratio?

Solution. Transmissibility for a free mass foundation is given by Eq. (4.13), with

$$m = 4.05 \times 10^4 \text{ kg,}$$

$$k = \frac{0.2025 \times 10^8}{0.198} = 1.023 \times 10^8 \text{ N/m}$$

$$\omega = 2\pi * 32 = 201.1 \text{ rad/s}$$

$$M = 200 * 10^3 = 2.0 \times 10^5 \text{ kg}$$

Thus, neglecting damping, transmissibility is given by

$$TR = \left| \frac{1}{1 - 16.0 + 0.2} \right| = 0.0676 = 6.76\%$$

This may be noted to be only marginally higher than that for rigid foundation (6.67%). So, compliance of the foundation is of no significance in this case.

Now, FR is given by Eq. (4.16), where

$$Z_1 = j\omega m, \quad Z_f = j\omega M \quad \text{and} \quad Z_2 = k/j\omega$$

Thus,

$$FR = \left| \frac{m + M}{m + M(1 - \omega^2 m/k)} \right| = \left| \frac{4.05 + 20}{4.05 + 20(1 - 16)} \right| = 0.081$$

This represents the real effectiveness of the set of isolator springs.

4.3.3 *Pneumatic suspension*

It has been shown before that for low transmissibility the natural frequency of the system should be much smaller than the forcing frequency. However, use of springs would result in large static deflection and mechanical instability. Therefore, low-frequency isolation is not feasible with metallic springs or viscoelastic pads. Fortunately, pneumatic suspension does not suffer from this weakness; it can yield good isolation with little static deflection even if the forcing frequency tended to zero!

Making use of the adiabatic gas law, it can be shown that stiffness of an air spring (a piston with a surge tank (see Fig. 4.6 for a basic configuration)), stiffness is given by [1]

$$\text{Single acting: } k \approx \gamma A^2 p_0 / V_0 \qquad (4.17)$$

$$\text{Double acting: } k \approx \gamma A^2 \left(\frac{p_1}{V_1} + \frac{p_2}{V_2} \right) \qquad (4.18)$$

where A is the load-bearing area, p_0 is the equilibrium pressure and V_0 is volume of the air in the cylinder and surge tank of a single-acting air spring shown in Fig. 4.6(a); p_1, p_2 and V_1, V_2 are the corresponding pressures and volumes of double acting spring shown in Fig. 4.6(b).

Force transmissibility or motion transmissibility (see Fig. 4.7) for rigid foundation can be obtained by means of Eq. (4.12) making use of the air spring stiffness from Eq. (4.18).

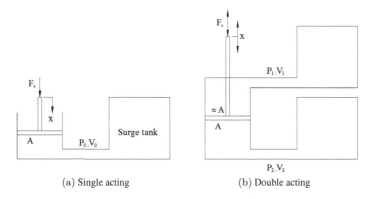

(a) Single acting (b) Double acting

Fig. 4.6   Schematic of pneumatic suspension (adapted from Ref. [1]).

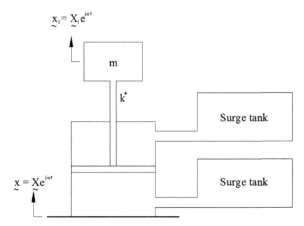

Fig. 4.7   Motion transmissibility of a double-acting air spring (adopted from Ref. [1]).

Damping can be introduced into the pneumatic suspension system through capillary passages connecting the surge tanks with the cylinder. For analysis of such a system, the reader is referred to Mallik [1].

The stiffness expressions (4.17) and (4.18) hold the key for designer who can select appropriate values of A, p_0 and V_0 to obtain the required (low enough) stiffness and adequate load-bearing capacity (Ap_0). Obviously, use of a large surge tank mould ensures low stiffness without compromising the load-bearing capacity. But then, there would invariably be a practical constraint on space availability.

4.4 Dynamic Vibration Absorber

A spring-mass (and damper) pair attached to a machine excited at or near its natural frequency as a mechanical appendage so as to absorb its energy is called a dynamic vibration absorber (DVA). Schematic of such an absorber is shown in Fig. 4.8(a). It was invented by Frahm at the beginning of the twentieth century, i.e., about 100 years ago. Its principle may be better understood by making use of analogous circuit [3] shown in Fig. 4.8(b). Here, m is the mass of the vibrating machine (element 2), k is the stiffness of the existing isolator spring (element 1), m_a is the DVA mass (element 4), and k_a is stiffness of the

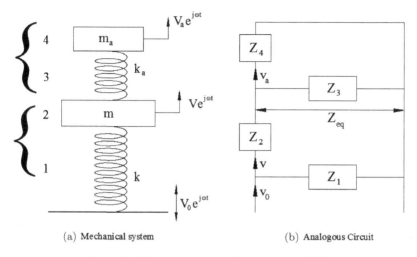

(a) Mechanical system   (b) Analogous Circuit

Fig. 4.8   Illustration of the function of a DVA.

DVA spring (element 3). For convenience, Fig. 4.8 shows velocities instead of displacements. To recall the analogous relationships [4].

$$v = j\omega x, \quad Z_1 = k/j\omega, \quad Z_2 = j\omega m, \quad Z_3 = k_a/j\omega, \quad Z_4 = j\omega m_a$$

$$(4.19)$$

Subscript a denotes absorber, or DVA.

In Fig. 4.8(b), Z_3 and Z_4 are in parallel. Therefore Z_{eq}, the equivalent impedance of the DVA on the machine mass m is given by

$$Z_{eq} = \frac{Z_3 Z_4}{Z_3 + Z_4} = \frac{\frac{k_a}{j\omega} \cdot j\omega m_a}{\frac{k_a}{j\omega} + j\omega m_a} = \frac{j\omega m_a}{1 - \omega^2 m_a/k_a} \qquad (4.20)$$

Clearly, if $k_a/m_a = \omega^2$, i.e., if the natural frequency of the absorber is equal to the forcing frequency, then $Z_{eq} \to \infty$. This would open up the lower loop in Fig. 4.8(b), making v, velocity of the machine mass m, tend to zero. This is the principle underlying the dynamic vibration absorber.

At the tuned frequency, $\omega = (k_a/m_a)^{1/2}$, the machine (mass m) would not move ($v = 0$). It would be subjected to a dynamic force $Z_1 v_0$ from below (the support side) and an equal and opposite force

$Z_3 v_a$ from above (see Fig. 4.8). Thus,

$$Z_3 v_a = Z_1 v_0 \qquad (4.21)$$

$$v_a = \frac{Z_1}{Z_3} v_0 = \frac{k}{k_a} v_0 = \frac{k}{m_a \omega^2} v_0 \qquad (4.22)$$

Thus, DVA mass m_a would have considerable motion (velocity v_a). In order to keep the DVA motion within limits, its mass m_a should not be too small.

A tuned vibration absorber ($\omega = (k_a/m_a)^{1/2}$) can also be used to reduce transmissibility of the original system, $|v/v_0|$, at frequency ratio, $r(\equiv \omega/\omega_n, \omega_n = (k/m)^{1/2})$ for $r > 1$. For this purpose, it can be proved that the effective range of a dynamic vibration absorber around its tuned frequency is proportional to the mass of the absorber and is governed by the approximate expression [3]:

$$\frac{\omega_2 - \omega_1}{\omega_a} = \frac{r_2 - r_1}{r_a} \simeq \frac{m_a}{m} \frac{1}{e}, \quad r_a = \frac{\omega_a}{\omega_n} = \frac{\left(\frac{k_a}{m_a} \right)^{1/2}}{\left(\frac{k}{m} \right)^{1/2}} \qquad (4.23)$$

where e is an arbitrary large number representing the designed isolation. Thus, if one wishes to have a transmissibility equal to one-fifth that of the corresponding straight-through system (without DVA) over a 5% variation in the forcing frequency r_a of 6, then,

$$e = \frac{1}{1/5} = 5$$

$$\frac{m_a}{m} \simeq \frac{r_2 - r_1}{r_a} \cdot e = \frac{5}{100} \times 5 = 0.25$$

$$\frac{k_a/m_a}{k/m} = (6)^2 = 36$$

Thus, Eqs. (4.23) adjust m_a and k_a in terms of the given m and k, completing thereby the design of the DVA.

Often, a DVA is designed to control the resonant vibration of a given machine or rotor or a sub-assembly. Then, $\omega = \omega_n = (k/m)^{1/2}$, and Eq. (4.22) reduces to

$$v_a = \frac{m}{m_a} v_0 \qquad (4.24)$$

Obviously, m_a must be large enough to limit its motion; otherwise, we would end up transferring the vibration problem from the machine to the DVA mass.

Figure 4.8 illustrates the use of a DVA for reducing the velocity transmissibility (v/v_0) of a sensitive instrument of mass m. However, as shown in Ref. [3], the same principle as well as Eq. (4.23) hold for reduction of the force transmissibility and amplitude of vibration of mass m acted upon by an oscillating force.

Example 4.4. Design an undamped dynamic vibration absorber (evaluate m_a and k_a) in order to reduce velocity transmissibility to one-fifth of the original value of 20% for a precision instrument weighing 10 kg mounted by means of a spring (see Fig. 4.8) on a surface vibrating vertically at about 100 Hz, with possible variation of 5%. Also, evaluate stiffness k of the main spring.

Solution. For the original (main) system,

$$TR = \left| \frac{1}{1 - r^2} \right| = 0.2$$

whence,

$$r = 2.45$$

or

$$\omega/\omega_n = f/f_n = 2.45 \Rightarrow f_n = \frac{f}{2.45} = \frac{100}{2.45} = 40.82 \text{ Hz}$$

Now,

$$\frac{1}{2\pi} \left(\frac{k}{m} \right)^{1/2} = f_n$$

or

$$\frac{1}{2\pi} \left(\frac{k}{10} \right)^{1/2} = 40.82 \Rightarrow k = (2\pi * 40.82)^2 * 10 = 6.58 * 10^5 \text{ N/m}$$

Besides, Eq. (4.23) gives,

$$\frac{\omega_2 - \omega_1}{\omega_a} = \frac{m_a}{m} \frac{1}{e}, \quad e = \frac{1}{1/5} = 5$$

Thus,

$$0.05 = \frac{m_a}{10} \cdot \frac{1}{5} \Rightarrow m_a = 2.5 \text{ kg}$$

A tuned dynamic vibration absorber would have its natural frequency equal to the forcing frequency. Thus,

$$\frac{1}{2\pi} \left(\frac{k_a}{m_a} \right)^{1/2} = 100 \Rightarrow k_a = (2\pi * 100)^2 * 2.5 = 9.87 \times 10^5 \text{ N/m}$$

Thus, a dynamic vibration absorber (DVA) with $m_a = 2.5$ kg and $k_a = 9.87 \times 10^5$ N/m would reduce the velocity transmissibility from 0.2 to $0.2 \times 0.2 = 0.04 = 4\%$. In other words, such a DVA would reduce amplitude of vibration of the instrument to 4% of that of the vibrating support.

One should use damping in parallel with spring or combine the two elements into a viscoelastic element. As a matter of fact, most DVAs incorporate damping in one way or the other. Design of optimally damped DVAs may be found in several textbooks as well as research papers — see, for example, Refs. [1, 5].

Figure 4.8 is just a schematic of a dynamic vibration absorber or vibration neutralizer or tuned mass damper. In practice, the primary vibrating system and the DVA may take a variety of shapes and forms. Some of the well-known applications are [1, 5]:

(a) the so-called stock bridge damper, widely used to reduce wind-induced vibration in the overhead power transmission lines,
(b) absorber for high-rise buildings, for suppressing primarily the contribution of the first vibration mode in wind-induced oscillations,
(c) pendulum-like DVA's applied to high television towers,
(d) devices used to:
   (i) stabilize ship roll motion,
   (ii) attenuate vibrations transmitted from the main rotor to the cockpit of helicopters,
   (iii) improve machine tool operation conditions,
   (iv) reduce the dynamic forces transmitted to an aircraft due to high rates of fire imposed on the canon motion,

(v) control torsional vibration of internal combustion engines and other rotating systems,

(vi) attenuate ship roll motion (gyroscopic DVA),

(e) centrifugal pendulum vibration absorber.

This subject is not discussed further here because the scope of this textbook is limited to control of vibrations that would lead to noise radiation.

4.5 Impedance Mismatch to Block Transmission of Vibration

It has often been observed that the unbalanced forces transmitted to the foundation or support structure travel long distance through the structure as structure-borne sound and radiate sound. This is also known as flanking transmission.

Therefore, reduction in noise radiation calls for:

(a) control of unbalanced forces and moments at the source,
(b) design of vibration isolators for minimal transmissibility,
(c) blocking transmission of vibration in the support structure by means of:

(i) impedance mismatch (reflection of power flux),
(ii) structural damping (absorption or dissipation of power flux).

4.5.1 *Viscoelastic interlayer*

Impedance mismatch is affected by sudden change in characteristic impedance along the transmission path. Common examples of the impedance mismatch are use of gaskets, washers, viscoelastic pads, etc., between two metallic layers or surfaces. The characteristic impedance Y (product of material mass density, ρ, and sound speed, c) of a viscoelastic material like rubber is several orders lower than that of a metal like steel.

It can be shown that the dynamic power transmission coefficient across an interface of two materials is given by [6]:

$$\tau = \frac{W_t}{W_i} = \frac{4Y_1Y_2}{(Y_1 + Y_2)^2}, \quad Y_1 = \rho_1 c_1, \quad Y_2 = \rho_2 c_2 \tag{4.25}$$

where W_t and W_i represent the incident power flux and transmitted power flux, respectively (see Fig. 4.9).

In order to appreciate that τ is always less than unity, we may rewrite τ of Eq. (4.25) as

$$\tau = 1 - \left(\frac{Y_1 - Y_2}{Y_1 + Y_2}\right)^2 \tag{4.26}$$

Thus,

$$\tau \ll 1 \quad \text{if } Y_1 \ll Y_2 \text{ and also if } Y_1 \gg Y_2 \tag{4.27}$$

In other words, transmission of dynamic power across an interface of two materials with vastly different characteristic impedances is very poor. This illustrates the principle and use of impedance mismatch.

Incidentally, symmetric nature of Eq. (4.26) shows that the power transmission coefficients for Figs. 4.9(a) and 4.9(b) would be the same.

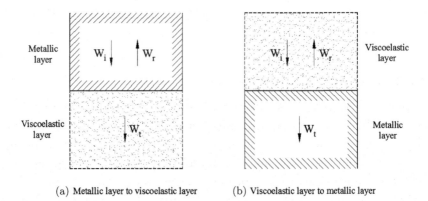

(a) Metallic layer to viscoelastic layer   (b) Viscoelastic layer to metallic layer

Fig. 4.9   Illustration of the principle of impedance mismatch.

Making use of the conservation of energy (or power flux) at the interface of two surfaces we can see that,

$$W_i - W_r = W_t \Rightarrow \frac{W_t}{W_i} = 1 - \frac{W_r}{W_i}, \quad \frac{W_r}{W_i} = \left(\frac{Y_1 - Y_2}{Y_1 + Y_2}\right)^2 \quad (4.28)$$

This shows that the effect of the impedance mismatch is to reflect a substantial part of the incident power back to the source. In other words, principle of impedance mismatch is the basis of reflective or reactive dynamical filters.

Figure 4.10 shows the use of a viscoelastic layer in between two metallic layers in order to block transmission of vibration (longitudinal waves, to be precise). This arrangement is very common in the mechanical engineering practice. It represents a pair of impedance mismatch interfaces: A–A and B–B. However, its overall transmission coefficient $\bar{\tau}$ is not as low as τ^2, where τ is the transmission coefficient of each of the two individual interfaces A–A and B–B. In fact, it would be nearly unity except for thick and very compliant interlayers and at high frequencies. This is because the vibrational energy reflected at the interface B–B, comes back substantially after

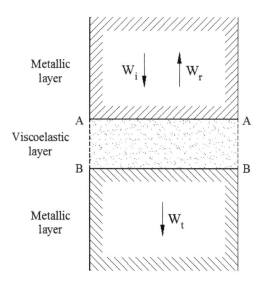

Fig. 4.10   Blockage of vibration transmission.

reflection from the interface A–A, and undergoes a series of back-and-forth reflections between the two interfaces before it gets absorbed or dissipated as heat within the viscoelastic layer. That is why, the intermediate layer must have viscosity (rather, structural damping) as well as elasticity.

Unlike τ of an individual interface, $\bar{\tau}$ is a function of frequency as well as thickness of the viscoelastic interlayer. In general, its effectiveness would improve ($\bar{\tau}$ would decrease) with increased thickness and higher forcing frequency.

4.5.2 *Effect of blocking mass on longitudinal waves*

Impedance mismatch for blocking longitudinal waves in beams (rods) and plates can also be created by means of a blocking mass, which represents a couple of changes in cross-section, similar to the two interfaces in the case of a viscoelastic interlayer in the foregoing subsection. It can be shown that the transmission coefficient of a blocking mass is given by [7]

$$\tau = \frac{1}{1 + \omega^2 M^2 / 4Y^2} \tag{4.29}$$

where ω is the forcing frequency of the longitudinal wave, M is the blocking mass (see Fig. 4.11), and Y is the characteristic impedance of the rod (or beam) for longitudinal wave propagation. It is given by [7]

$$Y = A'(\rho E')^{1/2} \tag{4.30}$$

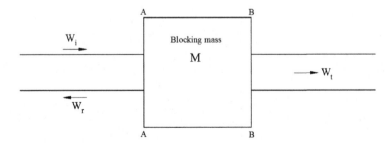

Fig. 4.11   Effect of blocking mass on vibration transmission.

where A' = cross-sectional area A for rods or beams, and thickness h for plates, E' = Young's Modulus E for beams and $E/(1-v^2)$ for plates, v and ρ are, respectively, Poisson's ratio and mass density of the material of the beam or plate.

A more common metric of the longitudinal wave transmission is Transmission Loss, TL. It is defined as

$$TL = 10\log(W_i/W_t) = 10\log(1/\tau) = -10\log(\tau), \text{ dB} \qquad (4.31)$$

TL of a blocking mass shown in Fig. 4.11 may be obtained by combining Eqs. (4.29) and (4.31). Thus,

$$TL = 10\log\left(\frac{1+\omega^2 M^2}{4Y^2}\right) \simeq 20\log\left(\frac{\omega M}{2Y}\right) = 20\log\left(\frac{|Z|}{2Y}\right) \qquad (4.32)$$

The approximate expression in Eq. (4.32) is for high frequencies and/or a large blocking mass such that $\omega M \gg 2Y$, and $Z = j\omega M$ is the inertive impedance of mass M.

Incidentally, the corresponding expression for the viscoelastic layer arrangement of Fig. 4.10 is given by

$$TL \simeq 20\log\left|1+\frac{Y}{2Z}\right| \simeq 10\log\left(\frac{1+\omega^2 Y^2}{4k^2}\right) \qquad (4.33)$$

where k is stiffness (inverse of compliance) of the viscoelastic layer; $k = EA/l$, E is Elastic modulus, A is area of cross-section, and l is thickness of the viscoelastic layer. If we wish to consider loss factor of the layer as well, then E and hence Z would be complex.

Example 4.5. A PVC gasket is used to isolate two steel surfaces. If the mass density and Young's modulus of the gasket are 1400 kg/m^3 and 2.4×10^9 N/m^2, respectively, and those of steel are 7800 kg/m^3 and 2.07×10^{11} N/m^2, respectively, then evaluate:

(a) characteristic impedance of the two materials,
(b) transmission coefficient and TL at each of the two interfaces for longitudinal waves, and
(c) TL due to the gasket of 3 cm thickness (see Fig. 4.10) at 10, 100 and 1000 Hz.

Solution. Characteristic impedance,

$$Y = \rho c = \rho \left(\frac{E}{\rho}\right)^{1/2} = (E\rho)^{1/2}$$

Y of steel,

$$Y_1 = (2.07 * 10^{11} * 7800)^{1/2} = 4.02 * 10^7 \ \text{kg}/(\text{m}^2\text{s})$$

Y of PVC,

$$Y_2 = (2.4 * 10^9 * 1400)^{1/2} = 1.83 * 10^6 \ \text{kg}/(\text{m}^2\text{s})$$

Use of Eq. (4.25) gives:

Transmission coefficient,

$$\tau = \frac{4Y_1Y_2}{(Y_1 + Y_2)^2} = \frac{4 * (4.02 * 10^7)(1.83 * 10^6)}{(4.02 * 10^7 + 1.83 * 10^6)^2}$$

$$= 0.166$$

Transmission loss,

$$TL = 10\log(1/0.166)$$

$$= 7.8 \ \text{dB}$$

Stiffness per unit area of the 3 cm gasket,

$$k = E/l$$

$$= \frac{2.4 * 10^9}{3/100}$$

$$= 8.0 \times 10^{10} \ \text{N}/\text{m}^3$$

Making use of Eq. (4.33), TL due to the gasket in Fig. 4.10 is given by

$$TL = 10\log\left\{1 + \frac{\omega^2 Y_1^2}{4k^2}\right\}, \quad \omega = 2\pi f$$

$$TL(f) = 10\log\left\{1 + \left(\frac{2\pi f * 4.02 * 10^7}{2 * 8.0 * 10^{10}}\right)^2\right\}$$

$$= 10\log\{1 + 2.49 * 10^{-6} f^2\}$$

Thus, $TL = 0.0$ at 10 Hz, 0.1 dB at 100 Hz, and 5.4 dB at 1000 Hz. It may note that TL of thin, soft inter-layers is negligible at lower frequencies. In other words, soft inter-layers would act as good filters only at higher frequencies.

The same expression (4.32) would yield TL of a blocking rotor of polar mass moment of inertia J for torsional shear-wave transmission along a circular torsional member-like shaft. Then,

$$Z = j\omega J \quad \text{and} \quad Y = I_p(\rho G)^{1/2}$$

where I_p is cross-sectional polar moment of inertia of the shaft, and G is the shear modulus of the shaft material.

Example 4.6. It is proposed to block the transmission of torsional shear waves along a steel shaft of 50 mm diameters by means of a rotor of 500 mm diameter, 100 mm thick, made of steel. Estimate the resulting TL for torsional waves at 125 Hz.

Solution. For steel,
Young's modulus, $E = 2.07 \times 10^{11}$ N/m^2
density, $\rho = 7800$ kg/m^3
Poisson's ratio, $v = 0.29$
Shear modulus is given by

$$G = \frac{E}{2(1+v)} = \frac{2.07 \times 10^{11}}{2(1+0.29)} = 0.8 \times 10^{11} \text{ N/m}^2$$

Cross-sectional polar moment of inertia,

$$I_p = \frac{\pi}{32}d^4 = \frac{\pi}{32} * (0.05)^4$$
$$= 6.136 \times 10^{-7} \text{ m}^4$$
$$Y_t = I_p(\rho G)^{1/2} = 6.13 * 10^{-7} * (7800 * 0.8 * 10^{11})^{1/2} = 15.313$$

Polar mass moment of inertia,

$$J = md^2/8 = \frac{\pi}{32} * (0.5)^4 * 7800 * 0.1 = 4.786 \text{ kg.m}^2$$
$$\omega = 2\pi f = 2\pi * 125 = 785.4 \text{ rad/s}$$

Finally, the torsional wave

$$TL = 10\log\left[1 + \frac{\omega^2 J^2}{4Y_t^2}\right]$$

$$= 10\log\left[1 + \left(\frac{785.4 * 4.786}{2 * 15.313}\right)^2\right]$$

$$= 41.8 \text{ dB}$$

4.5.3 *Effect of blocking mass on flexural waves*

Noise radiation from vibrating beams and plates is due to flexural or transverse vibration. Here, the motion is perpendicular to the beam (or plate) while the wave (or disturbance) moves along the axis of the beam. The governing equations of flexural waves in beams and plates, and the concepts of flexural wave number, propagating (far field) waves, evanescent (nearfield) waves, etc., were dealt with in Chapter 3 (see Section 3.6). Now, if there is a lumped blocking mass (line mass) M located at $x = 0$ on an infinite uniform beam (rather, in the centre of two semi-infinite beams), then the far field power transmission coefficient is given by [7]

$$\tau = \frac{(\bar{\mu}\bar{k} + 4)^2}{(\bar{\mu}\bar{k})^2 + (\bar{\mu}\bar{k} + 4)^2} \tag{4.34}$$

where

$\bar{k} = \sqrt{12}\,kr$ is a non-dimensional wave number,

$\bar{\mu} = \frac{M}{\sqrt{12\rho Ar}}$ is a non-dimensional blocking mass,

$r = (I/A)^{1/2}$ is the radius of gyration of the beam,

$k = \left(\frac{\rho A\omega^2}{EI}\right)^{1/4}$ is the flexural wave number, and

ρ, A and I are density, cross-sectional area and moment of inertia of the beam.

The corresponding relationships for plates with stiffeners may be found in references [8, 9]. Each stiffener is a lumped line mass. It not only provides additional stiffness against bending, but also reflects substantial portion of the flexural wave energy back to the source. Thus, stiffeners on hull plates of ships and submarines help in blocking of the flexural waves excited at the engine mounts or at the propeller thrust bearings, thereby reducing the noise emitted by the hull plate out into the sea or into the passenger cabins.

4.6 Damping Treatments for Plates

As indicated before, often the sheet metal components generate lot of noise. For example, the real source of noise in an internal combustion engine is the sharp pressure fluctuations in the cylinder due to combustion. However, most of noise radiation occurs from the sheet metal components like oil pan, valve covers, gear covers, the intake and exhaust manifolds, etc. These components act as mechanical loudspeakers. This combustion–excited body noise may be reduced either by making these components out of materials with high loss modulus, or by means of damping treatments. Almost all practical damping materials are polymeric plastics or elastomers [10].

As noise is radiated primarily by flexural vibration, damping treatments or viscoelastic laminae are used for absorbing the energy of flexural waves in thin plates.

Damping treatments are of two types:

(1) free or unconstrained layer damping (see Fig. 4.12(a)),
(2) sandwich or constrained layer damping (see Fig. 4.12(b)).

In free layer damping (FLD), the flexural wave energy is dissipated through the mechanism of extensional deformation, whereas in constrained layer damping (CLD), the wave energy is dissipated through shear deformation.

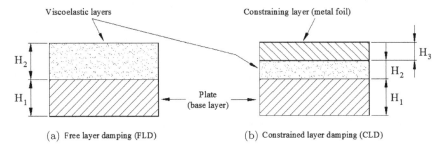

Fig. 4.12   Two types of damping treatments.

4.6.1 *Free layer damping treatment*

The FLD treatment often consists of a homogeneous, uniform adhesive layer of mastic deadener. Its overall loss factor is given by [1,11]

$$\eta \approx \frac{\eta_{E2} e_2 h_2 (3 + 6h_2 + 4h_2^2)}{1 + e_2 h_2 (3 + 6h_2 + 4h_2^2)} \tag{4.35}$$

where

η_{E2} = loss factor of the viscoelastic layer in longitudinal deformation,

$e_2 = E_2/E_1, h_2 = H_2/H_1, e_2 h_2 \ll 1$,

E_2 = storage modulus of the viscoelastic layer,

E_1 = Young's modulus of the base layer (plate),

H_2 = thickness of the viscoelastic layer,

H_1 = thickness of the base layer (plate).

The numerator of Eq. (4.35) indicates that in order to increase the overall loss factor, we need to increase $\eta_{E2} E_2$; i.e., the loss modulus of the viscoelastic layer, not its loss factor alone. Loss modulus of the best available commercial materials is of the order of 1 GPa $(= 10^9 \text{ N/m}^2)$.

For most of the useful range of applications, Eq. (4.35) may be replaced with [11],

$$\eta \approx 14 \eta_{E2} e_2 h_2^2 \tag{4.36}$$

η is less than $0.4\eta_{E2}$ for most practical FLD treatments unless we make use of a very thick treatment, $H_2 \gg H_1$ in Fig. 4.12(a). Normally $h_2 = H_2/H_1$ is of the order of unity, and then Eq. (4.35) yields the following order-of-magnitude relationship:

$$\eta \sim \eta_{E2} e_2 h_2^2 \sim 0.01 \eta_{E2} \tag{4.37}$$

Therefore, the FLD or unconstrained layer damping treatment leads to very little damping, and combined with its disadvantage of additional weight, it is rarely a cost-effective solution.

Example 4.7. A 2-mm thick aluminum plate is lined with a 2-mm thick viscoelastic layer of $E = 2 \times 10^9 (1 + j\, 0.5) \text{ N/m}^2$. Calculate the composite loss factor of the lined plate.

Solution. We make use of Eq. (4.35) for FLD.

For aluminum, storage modulus, $E_1 = 0.716 \times 10^{11}$ N/m^2.

We neglect its loss modulus; i.e., $\eta_1 = 0$.

Loss factor of the viscoelastic layer in longitudinal deformation, $\eta_2 = 0.5$.

Therefore,

$$e_2 = \frac{E_2}{E_1} = \frac{2 * 10^9}{0.716 * 10^{11}} = 0.028$$

$$h_2 = \frac{H_2}{H_1} = \frac{2 \text{ mm}}{2 \text{ mm}} = 1$$

Applying Eq. (4.35) for FLD, the composite loss factor is given by

$$\eta \approx \frac{0.5 * 0.028 * 1(3 + 6 * 1 + 4 * 1^2)}{1 + 0.028 * 1(3 + 6 * 1 + 4 * 1^2)} = \frac{0.5 * 0.364}{1 + 0.364} = 0.133$$

4.6.2 *Constrained layer damping treatment*

Typically, a CLD treatment shown in Fig. 4.12(b) is more effective than an FLD treatment shown in Fig. 4.12(a) in that the CLD treatment yields higher composite loss factor with lesser penalty in terms of additional weight. The shear deformation caused in a CLD beam is typically much more than the extensional deformation in an FLD beam. Therefore, the shear damping due to CLD treatment is much more than the extensional damping due to FLD treatment.

A CLD treatment often consists of a thin stiff metal foil with a thin adhesive damping layer on one of its sides. This can easily be stuck or pasted on to the base plate to be damped. The overall or composite loss factor of the arrangement shown in Fig. 4.12(b) is given by [1, 11]

$$\eta = \eta_{G2} Y g / [1 + (2 + Y)g + (1 + Y)(1 + \eta_{G2}^2)g^2] \qquad (4.38)$$

where

$$Y \approx 3.5 e_3 h_3 \text{ is the stiffness parameter,} \qquad (4.39)$$

η_{G_2} = loss factor of the viscoelastic material (layer 2) in shear,
$e_3 = E_3/E_1$, $h_3 = H_3/H_1$,
g, the shear parameter, is given by [11]

$$g = \frac{G_2}{k_b^2 H_2} \left[\frac{1}{E_1 H_1} + \frac{1}{E_3 H_3} \right] \tag{4.40}$$

G_2 = storage shear modulus of the viscoelastic layer,
k_b = the frequency-dependent flexural wave number of the composite structure, given by Eq. (2.53).

Equation (4.39) implies that the stiffer the constraining layer (metal foil), the higher the value of the stiffness parameter Y and thence the composite loss factor η.

Example 4.8. The 2-mm thick aluminum plate of Example 4.7 is now lined with a 1.0 mm thick aluminum foil with a thin (0.2 mm thick) adhesive damping layer on the inner side to provide constrained layer damping. If the shear modulus of the adhesive (damping) layer is $1.0 \times 10^7 (1 + j\,1.4)$ N/m^2, then evaluate the composite loss factor of the CLD plate at 25 Hz.

Solution. For aluminum, $E_1 = 0.7 * 10^{11}$ N/m^2 and $\rho_1 = 2700$ kg/m^3.

Let us neglect its loss factor; i.e $\eta_1 = 0$.

For the adhesive layer, $G_2 = 1.0 * 10^7$ N/m^2 and $\eta_{G2} = 1.4$ (given).

Circular frequency $\omega = 2\pi * 25 = 157.1$ rad/s.

$H_1 = 2$ mm, $H_3 = 1.0$ m and both layers are made of aluminum. Therefore, $h_3 = H_3/H_1 = 1.0/2 = 0.5$, and $e_3 = E_3/E_1 = 1.0$.

As per Eq. (4.39), stiffness parameter, $Y \approx 3.5 * 1.0 * 0.5 = 1.75$

Thickness of the damping layer, $H_3 = 0.2$ mm.

As per Eq. (2.53), the bending wave number is given by

$$k_b = \left(\frac{\rho_1 A_1 \omega^2}{E_1 I_1} \right)^{1/4} = \left[\frac{2700b * (2/1000)(157.1)^2}{0.7 * 10^{11} * b(2/1000)^3/12} \right]^{1/4}$$

$$= 7.3 \text{ m}^{-1}$$

Substituting these values in Eq. (4.40) yields the following value for shear parameter:

$$g = \frac{1.0 * 10^7}{(7.3)^2(0.2/1000)} \left[\frac{1}{0.7 * 10^{11} * (2/1000)} \right.$$

$$\left. + \frac{1}{0.7 * 10^{11} * (1.0/1000)} \right]$$

$$= 20.1$$

Finally, making use of Eq. (4.38) we evaluate the composite loss factor:

$$\eta = \frac{1.4 * 1.75 * 20.1}{1 + (2 + 1.75)20.1 + (1 + 1.75)(1 + 1.4^2)(20.1)^2}$$

Hence,

$$\eta = 0.0146$$

It may be noted that unlike the composite loss factor of FLD, which is independent of frequency, that of the CLD treatment is frequency dependent, even if properties (G and η) of the damping layer are constant (independent of frequency). Thus, the effectiveness of the CLD treatment depends on the mode of vibration at the forcing frequency. The composite loss factor, η, will be maximum if the CLD is applied at and near the node where shear strain is maximum.

The quantitative results presented above are strictly for a simply supported beam. Qualitatively, however, they apply for designing CLD treatments for plates and sheet-metal components as well. In particular, near the clamped or fixed edges where the flexural strain is the maximum, CLD treatment will be most effective. Unlike FLD treatment, the CLD treatment need not be applied over the entire exposed surface. A modal analysis would help in the selective or economic use of CLD treatment, particularly for narrow-band excitation.

In summary, the CLD treatment yields higher damping and imposes lesser penalty in terms of additional weight, as compared to the corresponding FLD treatment.

4.7 Active Vibration Control

An AVC system reduces the vibration of a system by means of an actuator that applies variable control input. When mechanical power is supplied to an AVC system the actuator is said to be a fully active actuator. Another type of actuator is the so-called semi-active actuator which is essentially a passive device which can store or dissipate energy. In this section, we restrict our discussion to the former; that is, a fully active actuator.

Active isolation actuators are generally of the following types:

(a) hydraulic actuators consisting of a reservoir, pump system and a fluidic circuit;
(b) electromagnetic drives like voice coil actuators;
(c) actuators incorporating active materials like piezoceramic, magneto strictive and magneto-rheological materials.

Piezoceramic materials tend to exhibit relatively high force and low stroke. Therefore, they are often combined with a hydraulic or mechanical load-coupling mechanism to multiply motion at the expense of the applied force [12].

Active vibration control is affected by one or a combination of the following approaches:

(a) feedback control,
(b) feedforward control.

In general, active isolation of periodic or otherwise predictable excitations is achieved by means of the feedforward control approach. However, if the excitation is random or unpredictable, then the feedback control approach is preferred. The two types of controllers exhibit exactly the same performance for the case of sinusoidal disturbance [12].

There is a lot of literature on the feedback control of large flexible systems at relatively low frequencies. However, feedforward control is the preferred approach for active control of vibration and the associated sound radiation at audio frequencies and steady-state excitation, which is of primary interest in this book. A block diagram of a feed-forward control arrangement is shown in Fig. 4.14. Here, all the signals are represented by Laplace transform and the system dynamics are represented by transfer functions.

As is obvious from Fig. 4.13, the response $E(s)$ is related to the excitation $X(s)$ as follows:

$$E(s) = X(s)[P(s) - H(s)]G(s) \qquad (4.41)$$

Here, $F_p(s)$ and $F_s(s)$ are, respectively, the primary force and secondary force acting on the mechanical system.

Unlike feedback control systems, the feedforward control arrangements require a high degree of accuracy in the magnitude and phase of the control system in order to obtain good cancellation. Therefore, most AVC systems make use of adaptive digital filters of the finite impulse response (FIR) type whose transfer function $H(s)$ is adapted via the error signal [13]. An example of such an AVC system is shown in Fig. 4.14 for control of flexural waves along a slender beam.

This type of AVC has been used, for example, to isolate vibrations from the gearbox traveling down the struts to the helicopter fuselage.

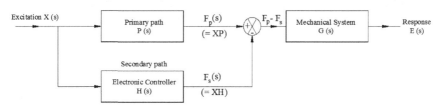

Fig. 4.13   Block diagram of a feedforward AVC system.

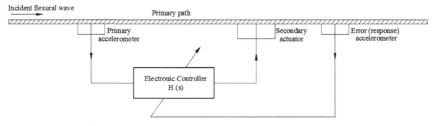

Fig. 4.14   Block diagram of an adaptive AVC system for control of flexural wave along a slender beam.

Active vibration control is particularly suited for engine mounts in an automobile. The disturbance vibrations that travel through the mounts to the chassis and thence to the passenger cabin of the vehicle are well correlated with the engine rotational speed. Therefore, the reference signal can be generated from the crankshaft pick-up. The error signal is obtained from an accelerometer mounted on the chassis and/or a microphone located inside the passenger cabin. The controller may be located at a convenient location like the boot or trunk of the vehicle. The active engine mount enables control inputs to be applied to the vehicle in the load path of the standard rubber passive mounts by means of control actuators co-located and integrated with the passive mounts. However, there being several paths in a vehicle for vibration transmission, the AVC system requires multiple control actuators and error transducers for global reduction of vibration in the structure and noise in the cabin [13].

For the same reason, global control of sound from vibrating structures through an AVC approach would require multiple actuators, multiple microphones as error transducers, and a multiple-input multiple-output (MIMO) controller. Here too, we can use adaptive FIR filters and the least mean square (LMS) algorithm to adapt the FIR filters. A prior knowledge of dominant modes would help in that the controller could be designed to control the efficient radiating modes [13]. Thus, a finite-element modal analysis of the structure is often an essential component of an active structural acoustical control (ASAC) system.

An important application of AVC is in active vibration absorbers (AVAs). Passive vibration absorbers suffer from the disadvantages of excessive stroke length and added mass. Piezo-electric materials like ceramic (PZT) or a polymer (PVDF) act as a transformer between mechanical and electrical energy. Thus, the AVA replaces the flexure and mass with electronic analogs and can achieve larger effective stroke length with less mass added to the system [14]. However, the AVAs too suffer from certain disadvantages as follows:

(a) AVA needs separate power and electronics and hence is substantially costlier than its passive counterpart.

(b) AVA needs custom analog circuits or digital controllers.

(c) Unlike mechanical (passive) vibration absorbers, AVAs can lead to dynamical instability [14].

References

[1] Mallik, A. K., *Principles of Vibration Control*, Affiliated East-West Press, New Delhi (1990).

[2] Lindley, A. L. G. and Bishop, R. E. D., Some recent research on the balancing of large flexible rotors, *Proceedings of the Institution of Mechanical Engineers*, 177, pp. 811–826 (1963).

[3] Munjal, M. L., A Rational synthesis of vibration isolators, *Journal of Sound and Vibration*, 39(2), pp. 247–265, (1975).

[4] Olson, M. F., *Dynamical Analogies*, Second Edition, Van Nostrand, Princeton (1958).

[5] Steffen, V. and Rade, D., Vibration absorbers, in *Encyclopedia of Vibration* (Ed. Braun, S. G.), Academic Press, San Diego (2002), pp. 9–26.

[6] Munjal, M. L., *Acoustics of Ducts and Mufflers*, Second Edition, Wiley, Chichester, UK (2014).

[7] Hayek, S. I., Vibration transmission, in *Encyclopedia of Vibration* (Ed. Braun, S. G.), Academic Press, San Diego, (2002), pp. 1522–1531.

[8] Cremer, L., Heckl, M. and Ungar, E. E., *Structure-borne Sound*, Springer-Verlag, New York (1973).

[9] Graff, K. F., *Wave Motion in Elastic Solids*, Ohio State University Press, Columbus, OH (1975).

[10] Ungar, E. E., Damping materials, in *Encyclopedia of Vibration* (Ed. Braun, S. G.), Academic Press, San Diego (2002), pp. 327–331.

[11] Ross, D., Ungar E. E. and Kerwin, E. M. (Jr.), Damping of plate flexural vibration by means of viscoelastic laminae, Section three, in *Structural Damping* (Ed. Ruzicks J. E.), Pergamon Press, Oxford (1960).

[12] Griffin, S. and Sciulli, D., Active isolation, in *Encyclopedia of Vibration* (Ed. Braun, S. G.), Academic Press, San Diego (2002), pp. 46–48.

[13] Fuller, C. R., Feedforward control of vibration, in *Encyclopedia of Vibration* (Ed. Braun, J. G.), Academic Press, San Diego (2002), pp. 513–520.

[14] Agnes, G., Active absorbers, in *Encyclopedia of Vibration* (Ed. Braun, S. G.), Academic Press, San Diego (2002), pp. 1–8.

Problems

4.1. A DG set running at 1500 RPM is to be isolated from the unbalanced forces by means of the AVM springs of Fig. 4.4 with stiffness of 912 N/mm (see Example 4.1). Masses of the engine, alternator and the base plate (inertia block) are 500, 250 and 2000 kg, respectively. How many of these AVM springs will be needed in order to limit transmissibility to 10%? Neglect the effect of damping.

[Ans.: 6]

4.2. The unbalance in the rotor of a reciprocating compressor is 100 gm-cm. The mass of the compressor is 500 kg, and it is operating at 1500 RPM, and is supported on rigid foundation through vibration isolators with static deflection of 10 mm. Calculate:

   (a) Amplitude of the unbalanced forces and the frequency of excitation.
   (b) Stiffness of the isolators and natural frequency of the system.
   (c) Transmissibility and amplitude of the force transmitted to the rigid foundation.
   (d) Transmissibility for a compliant foundation such that the static deflection is doubled.

[Ans.: (a) 24.67 N *at* 25 Hz; (b) 4.9 × 10⁵ $\frac{N}{m}$ and 4.98 Hz; (c) 0.041 and 1.02 N; (d) 0.02]

4.3. In order to arrest flanking transmission, an anechoic room, 5 m × 5 m × 5 m, is to be mounted on composite bonded rubber springs of the type shown in Fig. 4.4. All the six surfaces (floor, walls and ceiling) of the room are made of 200 mm thick RCC of density 2300 kg/m³. Neglecting the weight of the acoustic wedges, evaluate the number of 0.5 m × 0.5 m × 0.05 m rubber pads in each of the 9 equally spaced springs so as to ensure natural frequency of 8 Hz. Assume elastic modulus of rubber to be 10⁸ N/m².

[Ans.: 26]

4.4. A DG set, weighing 1000 kg and running at 3000 RPM, is supported on rigid foundation through eight AVMs of the type shown in Fig. 4.4, with $\alpha = 60^0$, $h = 5$ cm, $A = 50$ cm², damping ratio, $\zeta = 0.2$, $G = 3 \times 10^6$ N/m². Evaluate transmissibility of the system.

4.5. A machine weighing 1000 kg is supported on a rigid foundation by means of vibration isolators of stiffness 10^6 N/m^2. It is acted upon by unbalanced forces of $1000\cos(150t)$N. Evaluate amplitudes of vibration displacement of the machine and the force transmitted to the rigid foundation as it is, i.e., without any DVA, and then design an undamped DVA (evaluate k_a and m_a) in order to reduce vibration amplitude of the machine and the transmitted force to 25% of the original values for 2% variation in the forcing frequency.

[**Ans.:** $X = 0.046$ mm, $F_t = 46.5$ N, $m_a = 80$ kg and $k_a = 1.8 \times 10^6 \; \frac{\text{N}}{\text{m}}$]

4.6. It is proposed to block transmission of longitudinal waves through a 3 mm diameter steel rod by means of a blocking mass M, as shown in Fig. 4.11. What would be the required mass for transmission loss of 10 dB at 100 Hz? What would be the TL due to this blocking mass for flexural waves?

[**Ans.: 2.71 kg; 4.0 dB**]

Chapter 5

Sound Transmission Through Multiple Media

5.1 Wave Propagation Through a Single Interface

The sound wave interfacing with the medium boundary may create reflection, absorption, and transmission, as shown in Fig. 5.1.

The total incident power is equated with the summation of the reflected power (W_r), absorbed (W_α), and transmitted power (W_τ), respectively. This can be expressed in the equation form as Eq. (5.1).

$$W_i = W_r + W_\alpha + W_\tau \tag{5.1}$$

Dividing both sides of the equation by the incident power, Eq. (5.1) can be expressed in terms of the power reflection coefficient (R_w) absorption coefficient (α) and transmission coefficient (τ) as

$$\frac{W_r}{W_i} + \frac{W_\alpha}{W_i} + \frac{W_\tau}{W_i} = 1 \tag{5.2}$$

$$R_w + \alpha + \tau = 1 \tag{5.3}$$

All these coefficients are functions of frequency and the angle of incidence. The power absorption coefficient value can be varied from 0 to 1. Perfectly reflective material has zero absorption coefficient, and perfectly absorptive material would have a value of 1. The transmission coefficient is expressed in terms of the transmission loss (TL) (dB) as Eq. (5.4).

$$TL = 10 \log_{10} \left(\frac{1}{\tau} \right) \tag{5.4}$$

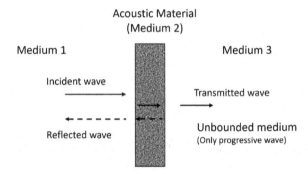

Fig. 5.1   Wave interaction at a medium boundary.

The transmission, reflection, and absorption coefficients can be defined in terms of power, intensity, and pressure. The definition of the Intensity transmission coefficient is the ratio of transmitted intensity to incident intensity [1]. It is written as

$$\tau = \frac{I_t}{I_i} = \left| \frac{p_t}{p_i} \right|^2 \tag{5.5}$$

Ideally, these definitions are given for normal plane waves. However, these definitions can be extended for all angles of incidence for calculating the reflection, absorption, and transmission coefficient. The expression for absorption coefficients in terms of pressure reflection coefficient for the no transmission conditions such as rigid backing plate to absorptive materials is

$$\alpha = 1 - |R|^2 \tag{5.6}$$

where R is the pressure reflection coefficient, defined as the ratio of the reflected pressure to the incident pressure.

The physics behind sound reflection, absorption, and transmission is explained with an ideal case of two unbounded mediums (like air and water) with a single interface, as shown in Fig. 5.2.

The wave penetration to medium 2 from medium 1 explains the absorption and transmission characteristics through a single interface. Only progressive waves are considered in medium 2 because it is unbounded. The density and speed of sound of the two mediums

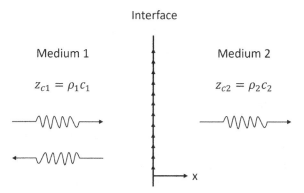

Fig. 5.2 Schematic diagram of two different unbounded mediums with a single interface.

are ρ_1, c_1, and ρ_2, and c_2, respectively. Characteristic impedances of the mediums are Z_{c1} and Z_{c2}, respectively.

At the interface, the impedance of the mediums changes, but the pressure and normal velocity to the surface are continuous.

For the plane wave incidence, the acoustic pressure in medium 1 at distance x from the arbitrary reference point is [2]

$$p_1(x,t) = (A_1 e^{-jk_1 x} + B_1 e^{jk_1 x})e^{j\omega t} \qquad (5.7)$$

where A_1 is incident wave amplitude in medium 1, B_1 is reflected wave amplitude and k_1 is the wave number, ω/c_1.

The particle velocity in the medium 1 is written as [2]

$$u_1(x,t) = \frac{(A_1 e^{-jk_1 x} - B_1 e^{jk_1 x})e^{j\omega t}}{Z_{c1}} \qquad (5.8)$$

The acoustic pressure in medium 2 is associated with penetration of the acoustic pressure from the interface, which is called transmitted pressure, and there is no reflected wave due to the unbounded medium. The progressive wave pressure in the medium 2 can be written as

$$p_2(x,t) = (A_2 e^{-jk_2 x})e^{j\omega t} \qquad (5.9)$$

The particle velocity in medium 2 can be written as

$$u_2(x,t) = \frac{(A_2 e^{-jk_2 x}) e^{j\omega t}}{Z_{c2}} \qquad (5.10)$$

where k_2 is the wave number, ω/c_2.

At the interface (at $x = 0$), applying continuity conditions leads to

$$p_1 = p_2|_{@x=0}, \quad u_1 = u_2|_{@x=0} \qquad (5.11)$$

Substitution of Eq. (5.11) in Eq. (5.8) yields to [2] pressure transmission coefficient,

$$\frac{A_2}{A_1} = \frac{2Z_{c2}}{Z_{c1} + Z_{c2}} \qquad (5.12)$$

and pressure reflection coefficient,

$$R = \frac{B_1}{A_1} = \frac{Z_{c2} - Z_{c1}}{Z_{c2} + Z_{c1}} \qquad (5.13)$$

Reflection is high for impedance mismatch, either from low impedance to high impedance (or) vice-versa. The reflected wave creates the standing wave, and pressure at the antinode decreases with the increasing absorption and limits the standing wave peak value. If the reflection coefficient is zero, then only a progressive wave exists.

The incident intensity amplitude in terms of incident pressure amplitude is [2]

$$I_i = \frac{|A_1|^2}{Z_{c1}} \qquad (5.14)$$

The reflected intensity amplitude in terms of reflected pressure amplitude is [2]

$$I_r = \frac{|B_1|^2}{Z_{c1}} \qquad (5.15)$$

The transmitted intensity amplitude in terms of transmitted pressure amplitude is [2]

$$I_t = \frac{|A_2|^2}{Z_{c2}} \qquad (5.16)$$

Ideally, these definitions are given for the normal-incidence plane waves. However, this definition can be extended for all angles of incidence for calculating the reflection, absorption, and transmission coefficient [3].

5.1.1 *Absorption in multilayer systems*

The absorption coefficient in terms of the pressure reflection coefficient can be written as

$$\alpha = 1 - |R|^2 = 1 - \left| \frac{Z_{c2} - Z_{c1}}{Z_{c2} + Z_{c1}} \right|^2 \tag{5.17}$$

Generally, the characteristic impedance can be complex for absorptive materials. Hence, the absolute values are calculated before squaring. The pressure, particle velocity amplitude, and phase may change when the wave strikes the surface. The pressure and particle velocity are in phase or 180° out of phase in a non-dispersive medium. This approach can be extended to calculate the complex amplitude and absorption coefficient for acoustic porous materials by modelling them as an equivalent fluid. A few cases are discussed hereunder on the medium impedance properties.

Case (i): $Z_{c2} = Z_{c1}$, there is no impedance mismatch between the two mediums. The sound wave propagates from one medium to other without change; and there is no reflection. In other words, absorption is complete.

Case (ii): $Z_{c2} > Z_{c1}$, wave propagation from a low impedance medium to high impedance medium, such as air to water medium. In this case, the wave sees the large surface impedance and cannot penetrate the second medium. This condition is like waves striking hard walls. At the rigid walls, particle velocity is zero, the reflected wave is in phase with the incident wave, and the standing wave pressure amplitude doubles near the surface.

Case (iii): $Z_{c2} < Z_{c1}$, wave propagation from a high impedance medium to low impedance medium, such as water to air medium. The reflected wave is 180° out of phase to the incident wave. So, the standing wave pressure at the interface is zero; this condition is termed the pressure-release surface. The absorption coefficient is minimal in both cases, and there is an impedance mismatch.

It is observed that impedance matching would provide maximum absorption. The absorptive materials dissipate the kinetic energy associated with particle velocity in terms of thermal and viscous losses. So, the absorptive material should be placed at the maximum particle velocity location.

5.2 Wave Propagation Through Multiple Interfaces

The propagating acoustic energy can be dissipated or blocked using multilayer absorbers or insulation configurations composed of perforated plates, airspaces, or porous materials, and supporting structures. Multilayer acoustical configurations are commonly used for various path control solutions like enclosures, acoustical treatment of the walls, dissipative mufflers, and plenums. The acoustical performance of these layers in terms of absorption coefficient and transmission coefficient or *TL* is explained here based on the wave equations for simple configurations.

The wave propagation formulation discussed for the single interface configurations in the previous section can be extended for multiple layer configurations by writing the pressure and particle velocity in terms of wave number and characteristics impedance of each medium and maintaining the continuity conditions at the layer interface. A simple configuration for multiple interfaces is shown in Fig. 5.3. Here, three materials have a characteristic impedance of Z_{c1}, Z_{c2}, and Z_{c3}, respectively. The acoustic pressure and particle velocity of each medium are written like Eqs. (5.4)–(5.6). The continuity conditions at the interface are pressure and velocity continuity at $x = 0$ and $x = L$. After simplifying equations, the pressure amplitude ratio for progressive waves leads to the reflection coefficient. The absorption coefficient is calculated from the reflection coefficient using Eq. (5.6).

It is observed from Fig. 5.4 that there are two-layer interfaces, such as air and material at $x = 0$ and the second interface between material and rigid wall at $x = L$. The particle velocity is zero at the

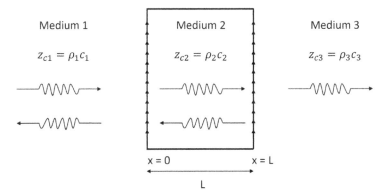

Fig. 5.3 Schematic diagram of three mediums with multiple interfaces.

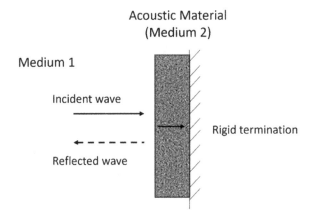

Fig. 5.4 Schematic diagram of acoustic layer with rigid backing.

rigid wall, and kinetic energy is small. The acoustical materials are backed with rigid walls in typical building acoustics applications, as shown in Fig. 5.4.

Hence, the placement of acoustic material with respect to the backing plate is critical to get the maximum absorption. The particle velocity is maximum at 0.25 times the wavelength from the rigid backing plate. Therefore, the acoustical treatment thickness should be more than 0.25 times the maximum wavelength for maximum absorption. The air gap is provided between the rigid wall and acoustic material for most practical applications. This air gap should be adjusted so that the material should see the impedance close to the air impedance.

5.3 Sound Transmission Through the Structure

The transmission coefficient in terms of medium impedance for single layer interface can be written by substituting Eq. (5.12) in Eq. (5.5).

$$\tau = \frac{I_t}{I_i} = \left|\frac{A_2}{A_1}\right|^2 \frac{Z_{c1}}{Z_{c2}} = \frac{4Z_{c1}Z_{c2}}{(Z_{c1} + Z_{c2})^2} \tag{5.18}$$

The TL of a single-layer interface configuration is calculated by substituting Eq. (5.18) into Eq. (5.4). It is a function of frequency and angle of incidence. The minimum TL value is 0 dB when impedances match and for acoustically transparent conditions (for example, door and window openings). Hence, sharp impedance mismatch provides the maximum TL. It is the basic concept used in various path control solutions, such as mufflers and partitions.

The transmission coefficient of multiple interface configurations shown in Fig. 5.3 can be written as follows:

$$\tau = \frac{I_t}{I_i} = \left|\frac{A_3}{A_1}\right|^2 \frac{Z_{c1}}{Z_{c3}} \tag{5.19}$$

where

$$\frac{A_1}{A_3} = \frac{1}{2}\left(1 + \frac{Z_{c1}}{Z_{c3}}\right)\cos(k_2 L) + j\frac{1}{2}\left(\frac{Z_{c1}}{Z_{c2}} + \frac{Z_{c2}}{Z_{c3}}\right)\sin(k_2 L) \tag{5.20}$$

Substituting Eq. (5.20) into Eq. (5.19) and simplification leads to the transmission coefficient as

$$\tau = \frac{4\frac{Z_{c1}}{Z_{c3}}}{\left(1 + \frac{Z_{c1}}{Z_{c3}}\right)^2 \cos^2 k_2 L + \left(\frac{Z_{c1}}{Z_{c2}} + \frac{Z_{c2}}{Z_{c3}}\right)^2 \sin^2(k_2 L)} \tag{5.21}$$

A few special cases that can be discussed for the multiple interfaces are as follows.

Case (i): If the medium 1 and medium 3 have the same impedance ($Z_{c1} = Z_{c3}$), then transmission coefficient can be simplified as

$$\tau = \frac{4}{4\cos^2 k_2 L + \left(\frac{Z_{c1}}{Z_{c2}} + \frac{Z_{c2}}{Z_{c1}}\right)^2 \sin^2(k_2 L)} \tag{5.22}$$

For small $k_2 L$, the trigonometric functions can be approximated as $\sin(k_2 L) \approx k_2 L$, $\cos k_2 L \approx 1$. The error in approximation is less than 3% for $k_2 L \leq 0.25$ rad.

Substituting the medium impedance and approximated trigonometric function values in Eq. (5.22) leads to

$$\tau = \frac{1}{1 + \left(\frac{Z_{c2}}{2Z_{c1}}\right)^2 (k_2 L)^2} \tag{5.23}$$

The specific mass (m_s) of medium 2 can be written as the density multiplied by the thickness. The transmission coefficient in terms of a specific mass, speed of sound, and density of the mediums can be written as

$$\frac{1}{\tau} = 1 + \left(\frac{\pi m_s f}{\rho_1 c_1}\right)^2 \tag{5.24}$$

The TL expression can be written as

$$TL = 10\log_{10}\frac{1}{\tau} = 10\log_{10}\left(1 + \left(\frac{\pi m_s f}{\rho_1 c_1}\right)^2\right) \tag{5.25}$$

The above TL equation is the mass law for panels for normal sound incidence. If the surrounding medium of the structure is air, then the above equation can be further simplified by substituting $\rho_1 c_1 = 400$ rayls as

$$TL = 20\log_{10}(f m_s) - K \tag{5.26}$$

where $K = 42$ dB for SI units.

The laboratory test conditions under a diffuse environment consider that the sound incidence angle is $0°$ to $90°$. However, the practical applications $0°$ to $72°$. Hence, the TL, defined as field TL, works out to be 5 dB lower than normal incidence TL. The constant K in Eq. (5.26) is 47 dB in mass law for the field TL calculations.

Case (ii): A special case can be considered when $k_2 L = n\pi$, $(n = 1, 2, 3, \dots)$

The trigonometric function values are $\cos^2 k_2 L \approx 1$, $\sin^2 k_2 L \approx 0$, the transmission coefficient given in Eq. (5.22) is simplified as

$$\tau = \frac{4 Z_{c1} Z_{c3}}{(Z_{c1} + Z_{c3})^2} \tag{5.27}$$

If the media 1 and 3 have the same characteristic impedance, then $\tau = 1$, or the $TL = 0$. It can be interpreted that medium 2 is behaving as acoustically transparent, and this occurs at the frequency of $f = \frac{n c_2}{2L}$.

The TL is measured as a function of frequency. However, the TL is expressed as a single number rating as sound transmission class (STC) for practical applications [1]. It is calculated from the one-third octave band data of TL from the band-centered frequencies of 125 to 4000 Hz. The measured TL data is compared with the reference curve as shown in Fig. 5.5. The slopes of the reference curve are as follows:

Frequency range (Hz)	Slope
125–400 Hz	9 dB/octave
400–1250 Hz	3 dB/octave
1250–4000 Hz	0 dB/octave (i.e., constant)

The reference curve is adjusted against the TL curve to satisfy the following two criteria: (i) the deviation at any band should not exceed 8 dB, and (ii) the total deviation in all the bands should not exceed 32 dB. The TL value at 500 Hz of the reference curve is STC. The subjective perception of the STC values is given in Table 5.1.

5.4 Acoustic Material Characteristics

Acoustic materials are characterized according to their macroscopic and microscopic properties. The macroscopic properties include the

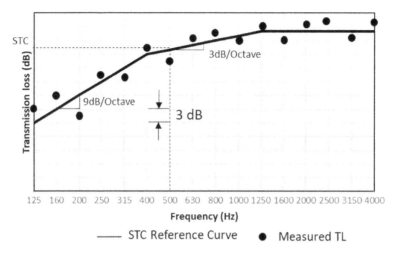

Fig. 5.5   Measured TL values superimposed on the standard STC contour.

Table 5.1.   Subjective perception of STC values.

STC	Subjective perception across the partition.
50–60	High amplitude sounds are heard faintly or not at all.
40–50	High amplitude sounds are heard faintly.
30–40	High amplitude sounds are heard but with lack of speech clarity.
20–30	Normal amplitude sounds are heard and understood clearly.

absorption coefficient and TL, whereas the microscopic properties include flow resistivity, porosity, tortuosity, viscous characteristic length, and thermal characteristic length [1, 3].

The sound absorption coefficient and TL are measured using an impedance tube (for normal incidence) and reverberation room technique (for random incidence). The details are given in Section 5.5.

Porosity (σ) is the percentage of open area within the material.

Tortuosity (T) measures the deviation of pores from a straight line through material. It is the ratio of the actual path length through the material to the linear path length, and the typical range is 1–10.

The airflow resistivity is the ratio of the pressure drop across a specimen to the linear velocity of airflow through a unit thickness of the specimen. It is given as

$$E = \frac{\Delta p}{ud} \qquad (5.28)$$

where E is the airflow resistivity in Pa-s/m^2, Δp is the pressure difference in Pa, u is the linear velocity in m/s, and d is the specimen thickness in m. Methods for measuring the flow resistivity can be categorized as the direct or steady airflow method, the alternating airflow method, the comparative method, and the acoustic method [4,5]. The flow resistivity is the combined effect of porosity, tortuosity, and material structure. The typical range of flow resistivity is 5000 to 40,000 Pa-s/m^2. In general, the material with higher flow resistivity provides better *TL*.

Another microscopic property that characterizes the porous material is the specific flow resistance. It is the ratio of the pressure drop across a specimen to the linear airflow velocity.

The dissipation mechanism in absorptive materials is due to the viscous and thermal losses in the pores of materials. These pores are in different sizes, and the average radius of smaller pores is related to viscous characteristic length. Similarly, the average radius of larger pores is related to thermal characteristic length. Generally, characteristic lengths are the ratio of the volume to surface area. Thermal characteristic length is higher than the viscous characteristic length.

The thermal characteristic length (Λ') equals twice the open porous network's volume to wet surface ratio [3].

$$\Lambda' = 2\frac{\oint dV}{\oint ds} \qquad (5.29)$$

The viscous characteristic length (Λ) is similar to this ratio weighted by the velocity in the volume and on the surface for an inviscid fluid [3]

$$\Lambda = 2\frac{\oint v^2 dV}{\oint v^2 ds} \qquad (5.30)$$

5.5 Measurement of Absorption Coefficient and Transmission Loss

Measurements are often necessary to verify whether a designed system or material satisfies design requirements. In the case of passive noise control, sound absorption coefficient and TL are the two main acoustic parameters of the material.

There are many standard methods available for the measurement of the above parameters.

A. Absorption coefficient

   (a) Normal incidence absorption — impedance tube method
   (b) Random incidence absorption — reverberation room method

B. Transmission loss

   (a) Normal incidence TL — impedance tube method
   (b) Random incidence absorption — SAE J1400 Method (two-room method)

Measurement of random incidence absorption coefficient and TL requires large-sized rooms [6, 7]. The normal absorption coefficient and TL [8] can be measured using an Impedance tube. The Impedance tube consists of a tube-like structure with an acoustic source (speaker) at one end and microphone holders along the tube for measurements and data acquisition systems. The other end is the termination of the impedance tube, which depends on the property to be measured. Figure 5.6 shows the two different arrangements of impedance tubes for absorption coefficient and TL measurements. The impedance tube method follows the international standards ASTM E2611-09 standards, ISO 10534-2, and ASTM E1050-08 [8, 9] for measuring absorption coefficient and TL. These are measured using the transfer function method (TFM). This method separates the incident pressure and the wave reflected sound pressure from the measured transfer function between two locations inside the tube, and then calculates the acoustic properties of the specimen installed in the tube.

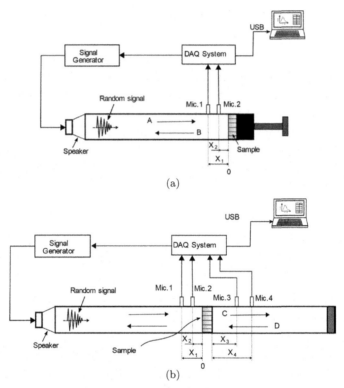

Fig. 5.6 Schematic diagram of impedance tube for measurement of (a) absorption coefficient and (b) *TL*.

The impedance tube setup consists of two different sized tubes, microphones, DAQ hardware, and measurement software. The different sizes of tubes are required to maintain plane-wave propagation along the tube for the frequency range of 63–6300 Hz, and the measured data can be combined later to obtain results over the complete frequency range. Laboratory images of the actual impedance tube measurement setup used to measure acoustic properties are shown in Fig. 5.7.

Impedance tube has the following components [7]:

- **Tubes:** Two different-sized tubes are used for different ranges of frequencies to measure the sample absorption coefficient and *TL*.

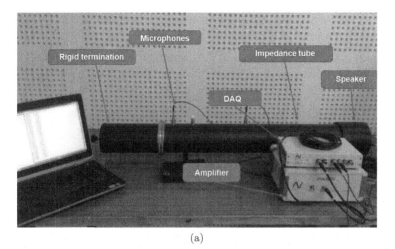

(a)

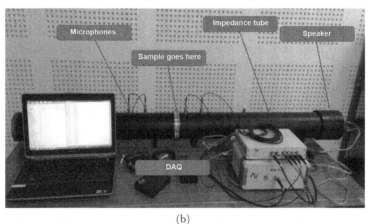

(b)

Fig. 5.7   Experimental setup of impedance tube for measurement of (a) absorption coefficient and (b) *TL*.

These tubes have internal diameters of 100 and 30 mm, with working frequency range of 63–1600 Hz and 800–6300 Hz, respectively. These sizes have been chosen to maintain plane wave propagation inside the tube. These tubes have a built-in speaker at one end and provision to install the microphones along the tube wall surface.

- **Microphones:** There are four numbers of 1/4″ microphones. These microphones have an excellent phase match as per standard requirements. These microphones fit into the holders provided on the tube and measure acoustic pressure inside the tube at that position.

- **DAQ:** Data-acquisition (DAQ) hardware has four channels for input and two channels for output. Microphones can be connected to the input channels to record sound pressures. The output channel is connected to the speaker through a power amplifier. The DAQ has a USB cable connection for the computer interface.

- **Power Amplifier:** The power amplifier provides an amplified signal to the speaker. A gain toggle adjusts the sound pressure level inside the tube between 90 and 110 dB. The output of the DAQ system connects to the input power amplifier, and its output connects to the speaker.

- **Calibrator:** It is used to calibrate microphones. It is a sound source with a 1 kHz tone of 94/114 dB SPL used for calibration. This calibrator has an adapter for the 1/4″ microphone. This calibrator follows the international standards IEC 60942:2003 Class 1, ANSI S1.40-1984, and GB/T 15173-1994.

5.5.1 *Measurement of the absorption coefficient*

The total sound power emitted by the speaker in the tube is the sum of acoustic power transmitted, absorbed, and reflected by the sample material [9]. The test sample is backed by the rigid termination in an impedance tube for absorption coefficient measurement. It leads to a very low transmission coefficient that can be neglected for further calculations. The impedance tube enables us to separate the incident and reflected sound components and thence, enables the estimation of the reflection coefficient. From the sound reflection coefficient, the absorption coefficient can be determined as follows [2]:

$$\alpha = 1 - |R|^2 \qquad (5.31)$$

Note that the R used in the above equation is the sound reflection coefficient calculated based on the amplitude of sound pressure, which is different from the sound power reflection coefficient.

Figure 5.8 shows the sample backed by rigid termination. In the figure, A is the pressure amplitude of the incident sound wave and B

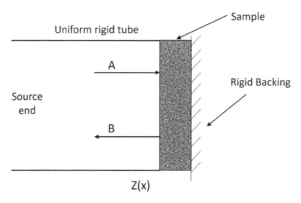

Fig. 5.8 Schematic of sample material backed by rigid termination in the impedance tube.

is the amplitude of the reflected sound. The impedance at the surface of the sample can be written as [2]

$$Z(x) = \frac{p(x,t)}{v(x,t)} = Z_c \frac{Ae^{-jkx} + Be^{jkx}}{Ae^{-jkx} - Be^{jkx}} \qquad (5.32)$$

or

$$Z(x) = Z_c \frac{e^{-jkx} + Re^{jkx}}{e^{-jkx} - Re^{jkx}} \qquad (5.33)$$

where R is the acoustic pressure reflection coefficient and is equal to the ratio of B to A. Here, Z_c is the characteristic impedance of the medium. From the above equation, the reflection coefficient can be determined by knowing the sample's impedance. During measurements, a test sample is mounted at one end of the impedance tube, a sound source at the upstream end generates a plane wave, and the sound pressures are measured at two locations near the sample (preferably less than three times the diameter of the tube).

The longest wavelength determines the distance between two microphones. The complex acoustic transfer function of the two microphone signals is determined and used to compute the normal-incidence complex reflection coefficient R, the normal-incidence absorption coefficient α, and the surface impedance of the test material Z.

5.5.2 *Calibration for phase factor and gain factor of transfer function*

Though a spectrum analyzer can easily measure the transfer function, it requires careful calibration. The use of an uncorrected transfer function in the calculation of material properties leads to wrong results. This correction can be done using the microphone switching method [7]. In this method, a transfer function is measured with the initial microphone configuration, and the second measurement is made by interchanging the positions of the microphones. But there is no change in the connection of the microphones with the data acquisition system. The geometric mean of the two results is taken as the desired result. Thus, the measurement errors due to phase mismatches between the microphones are eliminated, and the effect of unequal gain factor of the two microphone channels is also eliminated.

If H^{12} and H^{21} are the two transfer functions in the initial and reversed configurations, respectively, then the correction factor K_{12} can be calculated as geometric mean of the two transfer functions:

$$K_{12} = \sqrt{H^{12}H^{21}} \tag{5.34}$$

After calculating the correction factor, the corrected transfer function H'_{12} is determined by the following equation:

$$H'_{12} = \frac{H^{12}}{K_{12}} \tag{5.35}$$

This corrected transfer function is then used to determine the reflection and absorption coefficients.

5.5.3 *Measurement of transmission loss*

For *TL* measurement, the rigid backing given to the sample is replaced by a second tube as shown in Figs. 5.6(b) and 5.7(b). This configuration uses four microphones; hence, this method is sometimes referred to as the four-microphone method. The acoustic pressures and velocities at the four microphones are given by

$$p(x_1) = Ae^{-jkx_1} + Be^{jkx_1}; \quad p(x_2) = Ae^{-jkx_2} + Be^{jkx_2}$$

$$p(x_3) = Ce^{-jkx_3} + De^{jkx_3}; \quad p(x_4) = Ce^{-jkx_4} + De^{jkx_4}$$

$$v(x_1) = \frac{Ae^{-jkx_1} - Be^{jkx_1}}{Z_c}; \quad v(x_2) = \frac{Ae^{-jkx_2} - Be^{jkx_2}}{Z_c}$$

$$v(x_3) = \frac{Ce^{-jkx_3} - De^{jkx_3}}{Z_c}; \quad v(x_4) = \frac{Ce^{-jkx_4} - De^{jkx_4}}{Z_c} \quad (5.36)$$

where C and D are the coefficients like A and B, on the right side of the sample. The transfer matrix for the test sample of thickness d can be written as follows:

$$\begin{bmatrix} P \\ v \end{bmatrix}_{x=0} = \begin{bmatrix} T_{11} & T_{12} \\ T_{21} & T_{22} \end{bmatrix} \begin{bmatrix} P \\ v \end{bmatrix}_{x=d} \quad (5.37)$$

The transfer matrix compares the upstream and the downstream variables across the test sample. The transfer matrix is determined by measuring the acoustic pressure at the two ends of the sample, as shown in Fig. 5.6(b). Note that the transfer matrix evaluation consists of the determination of four unknowns. Hence, two more equations need to be generated by using two different terminations at the radiation end of the tube. Rigid termination and open-end termination represent a good pair for this two-load method of measurement. After calculating the transfer matrix, the TL is determined by the following equation:

$$TL = 20 \log_{10} \left| \frac{T_{11} + \frac{T_{12}}{Z_c} + T_{21} Z_c + T_{22}}{2} \right| \quad (5.38)$$

By measuring the acoustic pressure at the four microphone locations shown in Fig. 5.6(b) and using Eq. (5.35), the unknown pressure amplitude coefficients (A, B, C and D) can be estimated as follows [7]:

$$A = \sqrt{G_{rr}} \frac{j(H_{1r}e^{jkx_2} - H_{2r}e^{jkx_1})}{2\sin k(x_1 - x_2)}$$

$$B = \sqrt{G_{rr}} \frac{j(H_{2r}e^{-jkx_1} - H_{1r}e^{-jkx_2})}{2\sin k(x_1 - x_2)}$$

$$C = \sqrt{G_{rr}} \frac{j(H_{3r}e^{jkx_4} - H_{4r}e^{jkx_3})}{2\sin k(x_3 - x_4)}$$

$$D = \sqrt{G_{rr}} \frac{j(H_{4r}e^{-jkx_3} - H_{3r}e^{-jkx_4})}{2\sin k(x_3 - x_4)} \tag{5.39}$$

where H_{ir} is the frequency response function between the complex sound pressures, p_i, and the complex reference signal, (p_r). Here microphone 1 is taken as a reference. However, G_{rr} is the auto spectrum of the reference signal, r. The x_i is the distance of respective microphones from the front surface of the sample in the impedance tube, as shown in Fig. 5.6(b). By using these coefficients, the pressure and, subsequently, acoustic particle velocity can be estimated. Once the pressure and particle velocity are estimated, the surface impedance, absorption coefficient, and *TL* can be calculated using Eqs. (5.32) or (5.33), (5.31) and (5.38), respectively.

5.5.4 *Limitations of the impedance tube method*

As with all instruments, the impedance tube method also has some limitations.

1. Acoustic properties measured in an impedance tube are for normal incidence of the acoustic wave. In actual working cases, most of the time, the incident sound at an angle (oblique) or at random incidence.
2. A plane wave can be generated in an impedance tube only if the excitation frequency is below the cut-off frequency of the duct. This cut-off frequency is inversely proportional to the diameter of the tube. This condition limits the highest frequency up to which the sample can be tested accurately in an impedance tube.

3. The lower limit frequency that can be measured depends on the spacing between two microphones. Thus, different tubes are required for different frequency ranges.
4. Also, to carry out the measurements, the sample needs to be cut precisely in a circular shape; otherwise, it would significantly affect the measured results.

5.5.5 *Reverberation room method*

Reverberation time T_{60} is defined as the time required for the average SPL to decay by 60 dB from its initial value when the source is suddenly switched off. Reverberation chamber is used for finding the random incidence absorption coefficient. The absorption coefficient is measured using a relation involving reverberation time of the test chamber. According to the standard ASTM C423 and ISO 354, reverberation time is measured from pressure measurements in a standard diffusive environment. The sample absorption coefficient in terms of reverberation time of the reverberation room without (t_1) and with test sample (t_2) is given by [9]

$$\alpha_s = \frac{0.161\,V}{S_2}\left(\frac{1}{t_2} - \frac{1}{t_1}\right) \qquad (5.40)$$

where α_s is the random incidence absorption coefficient of the test sample, V is volume of the reverberation room, S_2 is surface area of a large enough test sample laid out in the middle of the floor of the reverberation room.

Figure 5.9 shows the schematic diagram of the reverberation time measurement setup. Generally, reverberation rooms are large (minimum volume of 150 m^3). So, as the distance traveled by sound rays increases, the uniformity of sound energy inside the room increases. All the surfaces consist of acoustically hard (reflective) material. There is a sharp change in characteristic impedance from air to the room's surface so that whatever sound energy falls on the surface, is reflected back into the room. None of the opposite surfaces are parallel, so standing waves won't be created inside the room. Sometimes, acoustic diffusers are placed at the ceiling of room, which help avoid standing waves and create a more effective diffusive field

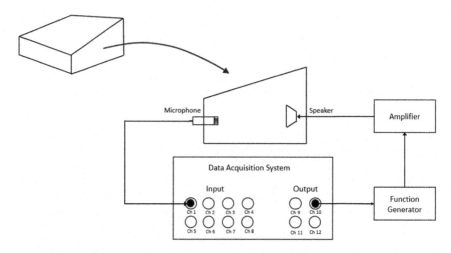

Fig. 5.9 Schematic diagram of the reverberation time measurement setup.

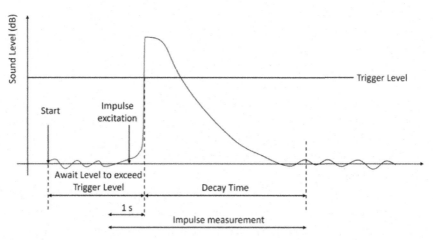

Fig. 5.10 Integrated impulse method for the reverberation time measurement [10].

environment. Reverberation time is a measure of the reverberation characteristics of the room.

There are two common methods to measure reverberation time, RT_{60}:

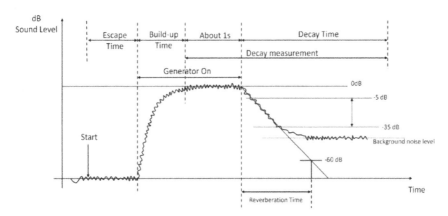

Fig. 5.11   Interrupted noise method for the reverberation time measurement [10].

(1) **Integrated impulse method:** Tone burst, or balloon burst, is given as input to the signal generator, and the reverberation time is measured from the decay curve. This process is carried out at different locations multiple times to get acoustic quality as a function of location. Reverberation time is obtained as a function of frequency. Figure 5.10 gives a typical decay curve for the integrated impulsive method.

(2) **Interrupted noise method:** A broadband noise (pink noise) is given to the source as input. For all frequency ranges of interest, it creates a constant sound pressure. Once a steady state is reached, the sound source is stopped suddenly, and from the decay, reverberation time is measured by extrapolating the curve. This process is carried out at several locations, multiple times, and a spatial average is taken to get the correct reverberation time of the room. Figure 5.11 shows a typical decay curve for the interrupted noise method.

References

[1] Ver, I. L. and Beranek, L. L., *Noise & Vibration Control Engineering*, John Wiley & Sons, NY (2005).
[2] Munjal, M. L., *Acoustics of Ducts and Mufflers*, Second Edition, Wiley, Chichester, UK (2014).
[3] Allard, J. F. and Noureddine, A., *Propagation of Sound in Porous Media: Modelling Sound Absorbing Materials*, John Wiley & Sons, NY (2009).

[4] Akiwate, D. C., Acoustic Analysis of Additive Manufactured Multilayer Periodic Structures, PhD thesis, Indian Institute of Technology Hyderabad (2019).

[5] ASTM C522-03, 2009, Standard Test Method for Airflow Resistance of Acoustical Materials.

[6] ISO 9053, Acoustics — Materials for Acoustical Applications — Determination of Airflow Resistance (1991).

[7] ASTM E2611−09, Standard Test Method for Measurement of Normal Incidence Sound Transmission of Acoustical Materials Based on the Transfer Matrix Method 1, Annu. B. ASTM Stand (2011), pp. 1–14.

[8] ASTM C423-09a, Standard Test Method for Sound Absorption and Sound Absorption Coefficients by the Reverberation Room Method (2009), pp. 1–12.

[9] ISO 354:2003 — Acoustics — Measurement of Sound Absorption in a Reverberation Room.

[10] Hottinger, B., and Kjaer, User Manual of Hand-held Analyzer Types 2270 (2016).

Problems in Chapter 5

1. A transformer tank is designed with a steel wall of a thickness of 5 mm, density of 7800 kg/m^3, and speed of sound in steel 5000 m/s. Estimate the change in *TL* of tank filled with and without oil at 1 kHz. The transformer oil density is 900 kg/m^3, speed of sound in oil is 1500 m/s, air density is 1.2 kg/m^3, and speed of sound in air is 340 m/s.

   [Ans.: 20.3 dB]

2. The human outer ear can be approximated as a pipe with one end open and another closed. The length and diameter of the pipe (air canal) are 30 and 5 mm, respectively. Calculate the fundamental natural frequency of the outer ear from the surface impedance. Derive the necessary equations for impedance and natural frequencies. Speed of sound is 342 m/s.

   [Ans.: 2850 Hz]

3. Explain the following terms: (i) Mass law, (ii) STC, (iii) reflection coefficient, (iv) *TL*, (v) flow resistivity, (vi) tortuosity, (vii) specific flow resistance.

4. A person is inside a swimming pool, and his friend stands outside and tries to convey some information. But it is not effectively

communicated. Why is it happening? Explain with appropriate mathematical expressions in terms of transmission coefficient.

5. Prove the following statement mathematically, "To maximize absorption, we need to minimize impedance mismatch between the two mediums."

6. It is desired to increase the TL of a panel in the mass-controlled region by 7 dB. Find the necessary change in the thickness.

[Ans.: Increase the thickness by 2.24 times]

7. (a) Explain the working principle of the Sound Impedance tube with proper schematic diagrams.

(b) Why do designers use two different diameter tubes?

(c) How to find the tube setup's minimum and maximum operating frequency?

8. Prove mathematically that the backing plate of the impedance tube without a sample is acoustically rigid. (*Hint*: Wave propagation through different media.)

9. Derive the mass law to calculate panel TL.

Chapter 6

Acoustics of Rooms, Partitions, Enclosures and Barriers

6.1 Introduction

Automobiles, trains and aeroplanes operate in more or less open space. However, machines are often installed and operated in closed factories and sheds where reverberant or diffuse sound field coexists with the direct sound field, as shown in Fig. 6.1. The direct sound field is characterized by a spherical or hemispherical wave front and its sound pressure level or the intensity level is governed by the inverse square law, as discussed before in Chapter 1. The reverberant field, by contrast, is generated by multiple reflections from the room walls. In an ideal reverberant room (with zero or little absorption), the sound pressure field would be diffuse; a microphone moved around in such a room would show more or less the same SPL everywhere. In an anechoic room, on the other hand, there will be no reflection or reverberations at all. Such a room would support only the direct sound field, and therefore, an anechoic room is said to simulate a free field like in the open outdoors. All real rooms like living quarters, offices, factories and commercial establishments support direct field as well as diffuse field to varying degrees, depending upon the amount of acoustic absorption of the ceiling, walls, floor, windows and doors as well as the furniture, carpeting, curtains and human occupancy.

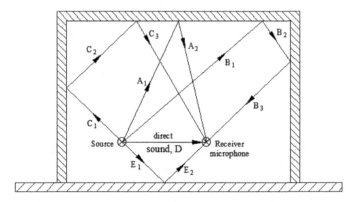

Fig. 6.1   Schematic of the direct sound field D superimposed on the reverberant sound field A_1–A_2, B_1–B_2–B_3, C_1–C_2–C_3 and E_1–E_2. The shortest distance between the source and the receiver microphone is denoted by r.

6.2 Sound Field in a Room

For a machine with a sound power level at a particular frequency, $L_w(f)$, the sound pressure level $L_p\,(r, f)$ at a distance r from the radiating surface or acoustical center of the machine (see Fig. 6.1) is, in general, given by the following formula [1]:

$$L_p(r, f) = L_w(f) + 10\log\left(\frac{Q(f)}{4\pi r^2} + \frac{4}{R(f)}\right) \qquad (6.1)$$

Here, $Q(f)/4\pi r^2$ is the contribution from the direct field, $4/R(f)$ is the contribution from the diffuse field and $Q(f)$ is the directivity factor.

$$Q(f) = Q_l \cdot Q_i(f) \qquad (6.2)$$

where Q_l is the locational directivity factor:

$$Q_l = 2^{n_s} \qquad (6.3)$$

n_s is the number of surfaces touching at the machine or source.
    Thus,

$Q_l = 1$ for a source suspended midair away from all surfaces $(n_s = 0)$;

2 for a source located in the middle of the floor, away from all walls ($n_s = 1$);

4 for a source located where the floor meets one of the walls ($n_s = 2$); and

8 for a source located in a corner where the floor meets two walls ($n_s = 3$).

$Q_i(f)$ is the inherent directivity factor of the source as a function of frequency.

Frequency f is often the center frequency of an octave band or one-third octave band. $R(f)$ is the room constant [2]:

$$R(f) = \frac{S\bar{\alpha}(f)}{1 - \bar{\alpha}(f)} \tag{6.4}$$

where S is the total surface area of the room, including floor, ceiling, walls, furniture and human occupancy, and $\bar{\alpha}(f)$ is the overall (surface-averaged) acoustic power absorption coefficient.

$$\bar{\alpha}(f) = \frac{\sum_i S_i \alpha_i(f)}{S}, \quad S = \sum_i S_i \tag{6.5}$$

where S_i and $\alpha_i(f)$ are the area of the ith surface, and its absorption coefficient, respectively.

Incidentally, it may be noted from Eq. (6.1) above that in free field where there is no reflection, $\bar{\alpha} = 1$, $R(f)$ tends to infinity, and then Eq. (6.1) yields the following direct field equation:

$$L_p(r, f)_{\text{direct field}} = L_w(f) + 10 \log \left(\frac{Q(f)}{4\pi r^2} \right) \tag{6.6}$$

where $Q(f)$ is given by Eq. (6.2) above.

In Eqs. (6.1) and (6.6) it is assumed that the microphone is in the far field. The criteria for the far field are complex [3]. For practical use, however, these may be simplified to

$$kr \geqslant 3, \quad r \geqslant 3l \tag{6.7a}$$

where $k = \omega/c = 2\pi/\lambda$ is the wave number, and l is the characteristic source dimension. For typical industrial sources of noise, the far-field criterion further simplifies to

$$r \geq 3l \tag{6.7b}$$

Equation (6.6) defines the inverse-square law for direct acoustic pressure field:

$$p_{rms}^2(r), \ I(r) \propto \frac{1}{r^2} \tag{6.8}$$

provided r is large enough to satisfy the far field criteria (6.7). In terms of levels, the inverse square law becomes,

$$L_p(r_1) - L_p(r_2) = 20 \log \left(\frac{r_2}{r_1} \right), \quad \text{dB} \tag{6.9}$$

As $20 \ \log 2 = 6$, Eq. (6.9) implies that SPL in the far field decreases by 6 dB per doubling of the distance. Thus, the decrease is sharper nearer to the source but is comparatively milder as one moves farther away from the source [1–3].

In a room, industrial shed or workshop, as we move away from the source, the direct field term of Eq. (6.1) decreases progressively as per the inverse square law, so that near to the walls, it becomes negligible with respect to the reverberant or diffuse field term $4/R$. Then Eq. (6.1) reduces to the diffuse field equation

$$L_p(f)_{\text{diffuse field}} = L_w(f) + 10 \log \left(\frac{4}{R(f)} \right) \tag{6.10}$$

Making use of Eq. (6.4) for the room constant R, Eq. (6.10) becomes

$$L_p(f)_{\text{diffuse field}} = L_w(f) + 10 \log \left[\frac{4(1 - \bar{\alpha}(f))}{S\bar{\alpha}(f)} \right] \tag{6.11}$$

Note that the diffuse field Eq. (6.11) is independent of the distance parameter r while the direct field Eq. (6.6) is independent of the room surfaces.

It may be noted from Eq. (6.11) that as $\bar{\alpha}(f)$ tends to zero, the diffuse field SPL would predominate over the direct field practically everywhere. This is the basic principle of a reverberation room. All the six surfaces of a reverberation room are designed to be highly reflective, and all three pairs of the opposite sides are constructed at an angle of 5 to 10 degrees so that the standing wave pattern is eliminated. This is how a reverberation room is designed to ensure a diffuse pressure field.

It may be noted from Eq. (6.1) that if r tends to zero, the direct sound field would predominate over the diffuse field. This is why a whisper in the ear can be heard even in the noisiest of environments. Thus, in a reverberant room sound level meter would show the same SPL everywhere except very near the source. So, the microphone should not be located too near the sound source while making measurements in a reverberation room.

The effectiveness of a reverberation room to produce a diffuse field is measured in terms of Reverberation Time, T_{60}, which is defined as the time required for the average sound pressure level to decay by 60 dB from its initial value (when the source is suddenly switched off). It is given by [2]:

$$T_{60} = \frac{55.25\,V}{Sc\bar{\alpha}} \approx 0.161\frac{V}{A} \qquad (6.12)$$

where V is the volume of the room, S is the total area of all surfaces of the room, c is sound speed, and A is the total absorption of the room in m^2.

$$A = S\bar{\alpha} = \sum_i S_i\alpha_i \qquad (6.13)$$

In practice, however, a room will have many complex objects, apart from its bounding surfaces (floor, ceiling, walls, windows, doors, and ventilators), and therefore, it is very difficult to evaluate the room absorption A from Eq. (6.13). Fortunately, it is now very easy to measure the reverberation time of the room all over the frequency range of interest. In fact, the random-incidence absorption coefficient for acoustic materials is now measured in a reverberation room as a standard practice [4,5]. The procedure in brief is as follows:

A blanket or mat of the test material is placed in the center of the floor in a reverberation room and the reverberation time is measured with and without the material. Then, repeated application of Eqs. (6.12) and (6.13) yields [4, 6]:

$$A \equiv S\alpha \approx 0.161V \left[\frac{1}{T_{60}} - \frac{1}{T'_{60}} \right] \qquad (6.14)$$

where T_{60} and T'_{60} are values of the reverberation time measured with and without the test material blanket or object, respectively, in the frequency band of interest. Thus, the absorption coefficient α of an acoustic material or absorption A of a test object can readily be evaluated experimentally.

Absorption coefficient of different types of walls, ceilings, furnishings, linings and panels at center frequencies of different octave bands, measured in specially designed reverberation rooms, are listed in handbooks — see, for example, references [1, 6, 7]. Some of them are reproduced here from Ref. [6] in Table 6.1.

As the pressure field in a reverberation room is more or less uniform throughout the room, Eq. (6.11) presents a convenient method of evaluating the total sound power level of a machine. Microphone at the end of the arm of a rotating boom is used to measure the space-average SPL in the frequency band of interest. Absorption A of the reverberation room is evaluated from the measured reverberation time and used in Eq. (6.11). This is in fact a standard technique for the measurement of sound power level of small portable machines for purposes of labeling as well as environmental impact assessment (EIA) [4, 8].

Example 6.1. The floor, walls and ceiling of a $15\,\mathrm{m} \times 10\,\mathrm{m} \times 5\,\mathrm{m}$ (high) room are made of varnished wood joists, bricks, and 13 mm suspended mineral tiles, respectively. Calculate the average absorption coefficient and reverberation time of the room in the 500 Hz frequency band when the floor is:

(a) unoccupied and unfurnished,
(b) unoccupied but furnished with well-upholstered seats,
(c) 100% occupied.

Table 6.1.   Sound absorption coefficient for common materials, objects and surfaces [6].

Material/surface/object	Octave band center frequency (Hz)					
	125	250	500	1000	2000	4000
Unoccupied average well-upholstered seating areas	0.44	0.60	0.77	0.84	0.82	0.70
Unoccupied metal or wood seats	0.15	0.19	0.22	0.39	0.38	0.30
100% occupied audience (orchestra and chorus areas) — upholstered seats	0.52	0.68	0.85	0.97	0.93	0.85
Audience, per person seated $S\bar{a}(\mathrm{m}^2)$	0.23	0.37	0.44	0.45	0.45	0.45
Wooden chairs — 100% occupied	0.60	0.74	0.88	0.96	0.93	0.85
Fibreglass or rockwool blanket $24\,\mathrm{kg/m}^3$, 50 mm thick	0.27	0.54	0.94	1.0	1.0	1.0
Fibreglass or rockwool blanket $24\,\mathrm{kg/m}^3$, 100 mm thick	0.46	1.0	1.0	1.0	1.0	1.0
Fibreglass or rockwool blanket $48\,\mathrm{kg/m}^3$, 100 mm thick	0.65	1.0	1.0	1.0	1.0	1.0
Polyurethane foam, $27\,\mathrm{kg/m}^3$, 15 mm thick	0.08	0.22	0.55	0.70	0.85	0.75
Concrete or terrazzo floor	0.01	0.01	0.01	0.02	0.02	0.02
Varnished wood joist floor	0.15	0.12	0.10	0.07	0.06	0.07
Carpet, 5 mm thick, on hard floor	0.02	0.03	0.05	0.10	0.30	0.50
Glazed tile/marble	0.01	0.01	0.01	0.01	0.02	0.02
Hard surfaces (brick walls, plaster, hard floors, etc.)	0.02	0.02	0.03	0.03	0.04	0.05
Gypsum board on $50 \times 100\,\mathrm{mm}$ studs	0.29	0.10	0.05	0.04	0.07	0.09
Solid timber door	0.14	0.10	0.06	0.08	0.10	0.10
13 mm mineral tile direct fixed to ceiling slab	0.10	0.25	0.70	0.85	0.70	0.60
13 mm mineral tile suspended 500 mm below ceiling	0.75	0.70	0.65	0.85	0.85	0.90
Light velour, $338\,\mathrm{g/m}^2$ curtain hung on the wall	0.03	0.04	0.11	0.17	0.24	0.35
hung in folds on wall	0.05	0.15	0.35	0.40	0.50	0.50
Heavy velour, $610\,\mathrm{g/m}^2$ curtain draped to half area	0.14	0.35	0.55	0.72	0.70	0.65
Ordinary window glass	0.35	0.25	0.18	0.12	0.07	0.04
Water (surface of pool)	0.01	0.01	0.01	0.015	0.02	0.03

Solution.

$$\text{Volume of the room} = 15 * 10 * 5 = 750 \, \text{m}^3$$
$$\text{Area of the floor} = 15 * 10 = 150 \, \text{m}^2$$
$$\text{Area of the ceiling} = 15 * 10 = 150 \, \text{m}^2$$
$$\text{Area of the walls} = 2(15 + 10) * 5 = 250 \, \text{m}^2$$

Total surface area of the room, $S = \Sigma S_i = 150 + 150 + 250 = 550 \, \text{m}^2$

Referring to Table 6.1, values of the sound absorption coefficient at 500 Hz for different surfaces are as follows:

Bare varnished wood joists floor: 0.1
Floor unoccupied but furnished with well upholstered seats: 0.77,
100% occupied floor: 0.85,
Brick walls: 0.03,
13 mm suspended mineral tile ceiling: 0.65.

Substituting these data on areas S_i and sound absorption coefficient α_i in Eq. (6.5) yields the following values of the surface-averaged $\bar{\alpha}$ at 500 Hz:

$$\bar{\alpha}(\text{bare room}) = \frac{150 * 0.1 + 250 * 0.03 + 150 * 0.65}{550} = 0.218$$

$$\bar{\alpha}(\text{upholstered seats}) = \frac{150 * 0.77 + 250 * 0.03 + 150 * 0.65}{550} = 0.4$$

$$\bar{\alpha}(\text{occupied room}) = \frac{150 * 0.85 + 250 * 0.03 + 150 * 0.65}{550} = 0.423$$

Substituting each of these three values of $\bar{\alpha}$ in Eq. (6.12) yields the corresponding values of the Reverberation Time, T_{60}. Thus,

$$T_{60}(\text{bare room}) = \frac{0.161 * 750}{550 * 0.218} = 1.0 \, \text{s}$$

$$T_{60}(\text{upholstered seats}) = \frac{0.161 * 750}{550 * 0.4} = 0.55 \, \text{s}$$

$$T_{60}(\text{occupied room}) = \frac{0.161 * 750}{550 * 0.423} = 0.52 \, \text{s}$$

Example 6.2. A noisy machine is located in the middle of the floor of the bare room in Example 6.1 ($S = 550\,\text{m}^2, \bar{\alpha} = 0.218$). Evaluate the distance (reckoned from the center of the machine) at which the direct field would be as strong as the diffuse field.

Solution.

$$L_p(r) = L_w + 10 \log \left(\frac{Q}{4\pi r^2} + \frac{4}{R} \right)$$

As the machine is touching one surface (the floor), $Q = 2^1 = 2$

Room constant, $R = \dfrac{S\bar{\alpha}}{1 - \bar{\alpha}} = \dfrac{550 * 0.218}{1 - 0.218} = 153.3\,\text{m}^2$

Contribution of the direct sound would be equal to the reverberation sound (diffuse field) if

$$\frac{Q}{4\pi r^2} = \frac{4}{R}$$

This would happen at

$$r = \frac{1}{4} \left(\frac{QR}{\pi} \right)^{1/2} = \frac{1}{4} \left(\frac{2 * 153.3}{\pi} \right)^{1/2} = 2.47\,\text{m}$$

Within a radius of 2.47 m, the direct sound would dominate, and outside this radius, the diffuse field would dominate.

Note that this neutral distance is a function of $\bar{\alpha}$, which in turn is a function of frequency (500 Hz in Example 6.1).

6.3 Acoustics of a Partition Wall

Often a noisy source in a room is separated from the receiver by means of a partition wall. When this wall does not extend upto the ceiling, it is called a barrier. Acoustics of a barrier is discussed later in this chapter. A partition wall reflects a part of the incident acoustic power or energy back to the source side of the room, absorbs a little of it and lets the rest pass through to the other side, as shown in Fig. 6.2. Then,

$$W_i = W_r + W_a + W_t \tag{6.15}$$

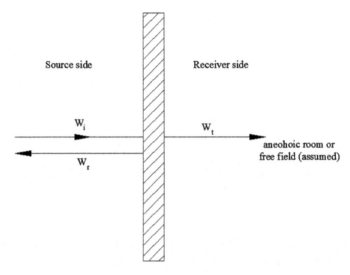

Fig. 6.2 Schematic of a partition wall for normal incidence transmission loss.

$$\text{Transmission coefficient, } \tau = \frac{W_t}{W_i} \qquad (6.16)$$

$$\text{Reflection coefficient, } R = \frac{W_r}{W_i} \qquad (6.17)$$

$$\text{Absorption coefficient, } \alpha = \frac{W_a}{W_i} \qquad (6.18)$$

$$\text{Transmission loss, } TL = 10 \log \left(\frac{W_i}{W_t} \right) = -10 \log(\tau) \qquad (6.19)$$

For most partition walls, absorption is negligible, i.e., $\alpha \ll 1$ unless the wall is lined with an acoustically absorptive layer. A major portion of the incident power is reflected back to the source side owing to the impedance mismatch between the air medium (on either side of the wall) and the material of the wall. For typical partition walls, lumped impedance of the wall, Z, normalized with respect to the characteristic impedance of the surrounding medium ($\rho_0 c_0$) (mostly, air) plays a primary role. It can be shown that transmission loss of

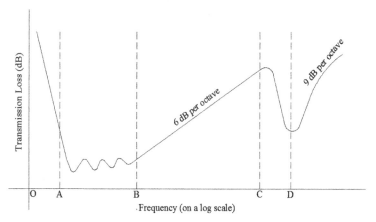

Fig. 6.3 Typical *TL* curve for a partition wall, showing the stiffness-controlled region, O–A, the damping-controlled region, A–B, the mass-controlled region, B–C, the coincidence (or, critical frequency) dip, D, and the coincidence-controlled region, beyond C.

a partition wall is given by

$$TL = 20 \log \left| 1 + \frac{Z}{2\rho_0 c_0} \right| \qquad (6.20)$$

Typically, impedance consists of a stiffness component, mass component and structural damping, and therefore a typical *TL* curve when plotted against frequency (on a log scale) has a stiffness-controlled region (O–A) and mass-controlled region (B–C) separated by a damping controlled region (A–B), as shown in Fig. 6.3. D denotes a Coincidence dip and the region beyond C is called the Coincidence region. The damping-controlled region is characterized by more than one resonance troughs, the lowest amplitude of which is controlled by structural damping of the partition wall. Depth of the coincidence dip, or *TL* at the critical frequency, is also controlled by structured damping.

For practical purposes, the mass-controlled region is most important, where,

$$Z \simeq j\omega m, \quad m = \rho h, \quad \omega = 2\pi f \qquad (6.21)$$

Here, m is surface density, defined as mass per unit area of the wall, ρ is density of the partition wall material, and h is thickness of the wall.

Substituting Eq. (6.21) in Eq. (6.20) yields the following formula for normal incidence TL:

$$TL = 10 \log \left\{ 1 + \left(\frac{\omega \rho h}{2 \rho_0 c_0} \right)^2 \right\} \simeq 20 \log \left\{ \frac{\pi f \rho h}{\rho_0 c_0} \right\} \text{ dB} \qquad (6.22)$$

Equation (6.22) applies for normal incidence. For random or field incidence we need to subtract 5.5 dB for one-third octave band measurements and 4 dB for octave band measurements.

It may be noted from Eq. (6.22) that in the mass-controlled region, TL increases by 6 dB per octave. Also, TL would increase by 6 dB if ρ, h or mass density m were doubled. In other words, thicker or denser partition walls would provide better acoustic insulation, and a given wall would provide more TL at higher frequencies; in fact, 6 dB additional TL per octave.

The phenomenon of coincidence occurs when the following relationship is satisfied for an incident wave striking an infinite wall or panel (see Fig. 6.4):

$$\lambda_b = \frac{\lambda_0}{\sin \theta}, \quad \lambda_0 = \frac{2\pi}{k_0} = \frac{c_0}{f} \qquad (6.23)$$

where λ_b is the bending or flexural wavelength,

$$\lambda_b = \frac{2\pi}{k_b}, \quad k_b = \left(\frac{m \omega^2}{EI} \right)^{1/4} \qquad (6.24)$$

m is the surface density of the wall,
EI is the flexural rigidity of the wall,
θ is the angle of incidence, and
k_b is the flexural wave number.

It is obvious from the coincidence relation (6.23) that coincidence would occur at only those frequencies that satisfy the inequality,

$$\lambda_b \geq \lambda_0 \qquad (6.25)$$

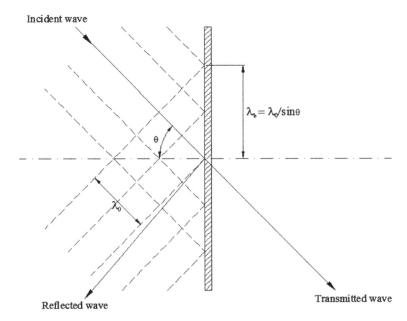

Fig. 6.4   Illustration of the phenomenon of coincidence.

This inequality would be satisfied at all frequencies exceeding the coincidence frequency given by

$$f_c = \frac{c_0^2}{2\pi} \left(\frac{m}{EI}\right)^{1/2} \qquad (6.26)$$

Above this critical frequency, wavelength of the flexural or bending wave in the wall will become equal to the projection of the wavelength of the incident wave upon the wall at an appropriate angle θ, as shown in Fig. 6.6.

Incidentally, Eq. (6.26) is particularly significant in as much as it represents the cut-off frequency below which a vibrating wall (or plate) would not radiate any sound, and conversely, an incident wave would not be able to set the wall (or plate) into vibration. This important physical principle is of great use in control of noise from vibrating bodies, as will become clear later in Chapter 6.

In practice, *TL* of partition walls, including panels, masonry walls, stud partitions, glazed windows, doors and floors, is measured in a reverberation suite for random incidence in different octave band center frequencies. For some common partition walls, the measured values of the random incidence transmission loss are listed in Table 6.2. These are extracted from Ref. [6].

It may be noted from Table 6.2 that the random-incidence TL of common structures and materials does not necessarily follow the shape of Fig. 6.3, not even the well-defined mass law. Nevertheless, Fig. 6.3 does serve as a useful guideline.

In particular, it is obvious that a double-glaze window with substantial airgap (≥ 50 mm) in between yields substantially more TL than a single-glaze window of double the thickness, because of two pairs of impedance mismatch in the case of a double-glaze window, or for that matter, double-skin door, double masonry walls, stud partitions, etc.

It may also be noted that it is advisable to fill up the whole or a part of the intervening airgap with an absorptive material.

Example 6.3. Making use of the mass law, calculate the normal incidence TL of a 16 g galvanized steel sheet at mid-frequencies of the octave bands of 63–8000 Hz, and compare the same with the corresponding random-incidence values from Table 6.2.

Solution.

Thickness of 16 g sheet, $h = 1.6$ mm

Mass density of steel, $\rho = 7800$ kg/m^3

Therefore, surface density of 16 g steel sheet,

$$M = \rho h = (1.6/1000) * 7800 = 12.48 \, \text{kg/m}^2$$

Characteristic impedance of air at 25°C, $\rho_0 c_0 = 410$ kg/(m^{2}s)

As per the mass law (Eq. 6.22), the normal-incidence TL of a sheet is given by

$$TL = 20 \log \left(\frac{\pi f \rho h}{\rho_0 c_0} \right) = 20 \log \left(\frac{\pi f * 7800 * 1.6}{410 * 1000} \right)$$

Table 6.2.   Measured values of the random-incidence TL of typical partition walls [6].

Panel construction	Thickness (mm)	Surface density (kg/m²)	Octave band center frequency (Hz)							
			63	125	250	500	1k	2k	4k	8k
Panel of sheet materials										
1.5 mm lead sheet	1.5	17	22	28	32	33	32	32	33	36
6 mm steel plate	6	50	—	27	35	41	39	39	46	—
22 g galvanized steel sheet	0.75	6	3	8	14	20	23	26	27	35
16 g galvanized steel sheet	1.6	13	9	14	21	27	32	37	43	42
Fibre board on wood framework	12	4	10	12	16	20	24	30	31	36
Plastic board sheets on wood framework	9	7	9	15	20	24	29	32	35	38
Woodwork slabs, unplastered	25	19	0	0	2	6	6	8	8	10
Woodwork slabs, plastered (12 mm on each face)	50	75	18	23	27	30	32	36	39	43
Plywood	6	3.5	—	17	15	20	24	28	27	—
Lead vinyl curtains	2	6.9	—	15	19	21	28	33	37	—
Single masonry walls										
Single leaf brick, plastered on both sides	125	240	30	36	37	40	46	54	57	59
Single leaf brick, plastered on both sides	255	480	34	41	45	48	56	65	69	72
Single leaf brick, plastered on both sides	360	720	36	44	43	49	57	66	70	72
Hollow cinder concrete blocks, painted	100	75	22	30	34	40	50	50	52	53
Hollow cinder concrete blocks, unpainted (cement base paint)	100	75	22	27	32	32	40	41	45	48
Aerated concrete blocks	100	50	—	34	35	30	37	45	50	—
Aerated concrete blocks	150	75	—	31	35	37	44	50	55	—

(*Continued*)

Table 6.2. (*Continued*)

Panel construction	Thickness (mm)	Surface density (kg/m²)	Octave band center frequency (Hz)							
			63	125	250	500	1k	2k	4k	8k
Stud partitions										
50 mm × 100 mm studs, 12 mm board both sides	125	19	12	16	22	28	38	50	52	55
50 mm × 100 mm studs, 9 mm plasterboard and 12 mm plaster coat both sides	142	60	20	25	28	34	47	39	50	56
Gypsum wall with steel studs and 16 mm-thick panels each side										
Empty cavity, 45 mm wide	75	26	—	20	28	36	41	40	47	—
Cavity, 45 mm wide, filled with fiberglass	75	30	—	27	39	46	43	47	52	—
Gypsum wall, 16 mm leaves, 200 mm cavity with no sound absorbing material and no studs	240	23	—	33	39	50	64	51	59	—
As above with 88 mm sound absorbing material	240	26	—	42	56	68	74	70	73	—
Single glazed windows										
Single glass in heavy frame	6	15	17	11	24	28	32	27	35	39
Single glass in heavy frame	8	20	18	18	25	31	32	28	36	39
Laminated glass	13	32	—	23	31	38	40	47	52	57

Doubled glazed window

2.44 mm panes, 7 mm cavity	12	15	15	22	16	20	29	31	27	30
9 mm panes in separate frames, 50 mm cavity	62	34	18	25	29	34	41	45	53	50
6 mm glass panes in separate frames, 100 mm cavity	112	34	20	28	32	38	45	45	53	50
6 mm and 8 mm glass, 100 mm cavity	115	40	—	35	47	53	55	50	55	—

Doors

Flush panel, hollow core, normal cracks as usually hung	43	9	1	12	13	14	16	18	24	26
Solid hardwood, normal cracks as usually hung	43	28	13	17	21	26	29	31	34	32
Typical proprietary "acoustic" door, double heavy sheet steel skin, absorbent in air space, and seals in heavy steel frame	100	—	37	36	39	44	49	54	57	60
Hardwood door	54	20	—	20	25	22	27	31	35	—

or

$$TL = -20.4 + 20 \log(f) \, \text{dB}$$

Values calculated from this relation are listed below in the second row.

Frequency (Hz)	63	125	250	500	1 k	2 k	4 k	8 K
Normal incidence, *TL* (as per mass law)	15.5	21.5	27.5	33.5	39.5	45.5	51.5	57.5
Random-incidence, *TL* (measured; Table 6.2)	9	14	21	27	32	37	43	42

It may be noted from the table that the normal-incidence mass law considerably overpredicts *TL* as compared to the measured values of *TL* for random incidence. This observation is very significant for designers of acoustic walls and enclosures. They should make use of the measured values of the random-incidence *TL* rather than the normal-incidence mass law.

6.4 Design of Acoustic Enclosures

Performance of an acoustic enclosure is measured in terms of its insertion loss. It is defined as reduction of SPL at the receiver due to location of the source (machine) in an acoustic enclosure. It is given by the following, rather approximate, relationship:

$$IL = 10 \log(\bar{\alpha}/\bar{\tau}) = TL + 10 \log \bar{\alpha} \tag{6.27}$$

Here, $\bar{\tau}$ and TL are the transmission coefficient and transmission loss of the impervious (often metallic) layer of the enclosure walls, and $\bar{\alpha}$ is the random-incidence absorption coefficient of the absorptive lining on the inner or source side of the enclosure walls (see Eq. (6.5)).

It may be noted from Eq. (6.27) that IL of the enclosure is always less than *TL* of the walls because α is always less than unity. It also indicates importance of the absorptive lining of all the inner surfaces of the acoustic enclosure. In the absence of the lining, the diffuse pressure field inside would be very high and therefore IL would be very small despite metallic walls of high surface density.

The surface-averaged transmission coefficient $\bar{\tau}$ is given by an equation similar to Eq. (6.5) for $\bar{\alpha}$. Thus,

$$\bar{\tau}(f) = \frac{\sum S_i \tau_i(f)}{S}, \quad S = \sum_i S_i \qquad (6.28)$$

As τ for openings or leaks is nearly unity, leaks could seriously compromise $\bar{\tau}$ and hence TL. For example, access openings or leaks of 1% would limit TL to 20 dB irrespective of the surface density of the walls of the enclosure. Similarly, access openings of 10% would limit TL to 10 dB. This indicates how ineffective could partial enclosures be, except at high frequencies where one could effectively make use of directivity of the high-frequency sounds. Partial enclosures would yield low IL values for low-frequency sounds, which tend to propagate spherically, easily skirting around the openings.

For practical purposes, Eq. (6.27) may be rewritten as

$$IL = TL - C, \quad C \equiv -10 \log \bar{\alpha} \qquad (6.29)$$

The experimentally determined values of C as functions of frequency for different types of enclosures are listed in Table 6.3 which is reproduced from Bies and Hansen [6].

Table 6.3. Values of constant C (dB) to account for enclosure internal conditions.

Enclosure internal acoustic conditions*	Octave band center frequency (Hz)							
	63	125	250	500	1000	2000	4000	8000
Live	18	16	15	14	12	12	12	12
Fairly live	13	12	11	12	12	12	12	12
Average	13	11	9	7	5	4	3	3
Dead	11	9	7	6	5	4	3	3

Note: *Use the following criteria to determine the appropriate acoustic conditions inside the enclosure:

Live: All enclosure surfaces and machine surfaces hard and rigid.
Fairly live: All surfaces generally hard but some panel construction (sheet metal or wood).
Average: Enclosure internal surfaces covered with sound-absorptive material, and machine Surfaces hard and rigid.
Dead: As for "Average", but machine surface mainly of panels.

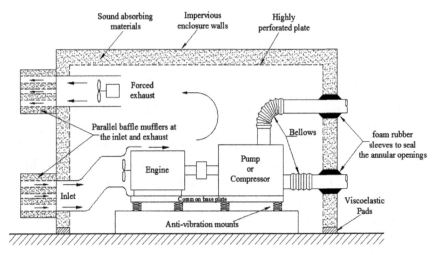

Fig. 6.5   Schematic of an acoustical enclosure used for noise control of an engine-driven pump or compressor [9].

In most application, the enclosure's internal surfaces are covered with a sound-absorptive material, whereas the machine surfaces are hard and rigid. Therefore, the 'average' row (the third row) of Table 6.3 would apply for constant C. Lining the machine surfaces with acoustic panels (see the 'dead' or the fourth row) gives only a marginal improvement in C and thence IL and is therefore not cost-effective.

Figure 6.5 shows schematic of an acoustic enclosure [9]. The following features of the acoustic enclosure may be noted:

1. All the available surfaces of walls and ceiling are lined with absorptive materials.
2. Parallel baffle mufflers are provided with sufficient openings for inlet and exhaust of cooling air.
3. The intake air is properly guided onto the hot surfaces of the machine. When ventilation for heat removal is required but the heat load is not large, then natural ventilation with silenced air

inlets low down close to the floor and silenced outlets at a greater height well above the floor, will be adequate.

4. Generally, forced ventilation is needed, and indeed provided, by means of booster fans at the inlet and/or exhaust points.

5. If forced ventilation is needed to avoid excessive heat buildup in the enclosure, then the approximate amount of airflow needed can be determined by $\rho C_p V = H/\Delta T$, where V is the volume flow rate of the cooling air required to limit the steady state temperature increase inside the enclosure to $\Delta T°C$, ρ and C_p are respectively the mass density and the specific heat at constant pressure of the cooling air, and H is the heat input to the enclosure in watts.

6. Flexible connectors are used to isolate the enclosure walls from machine vibrations.

7. Sealed openings are provided in the enclosure walls for any service pipes.

8. The noisy machine and its driver motor are mounted on a common platform, which rests on the foundation block through vibration isolators.

9. Viscoelastic pads support the enclosure all around in order to arrest flanking transmission of the structure-borne sound.

10. It is worth mentioning here that for a practical acoustic enclosure, TL in Eq. (6.27) is considerably more than the values given for a flat plate in Table 6.2. In practice, the enclosure is made of box-like rectangular frames, where the impervious layer (plate) gets considerably stiffened, and TL of a stiffened plate is more than that of a flat plate (of the same surface density) by several decibels. Some of this advantage may, however, be lost in acoustic leaks at the inter-frame joints if sufficient care is not exercised in assembling of the enclosure walls.

Figure 6.6 shows different types of penetrations (for cooling) lined with absorptive materials [9]. Theory and design of parallel baffle

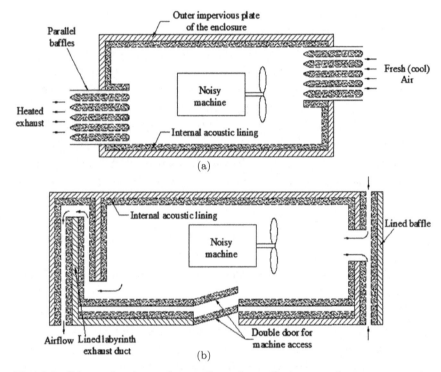

Fig. 6.6 Schematic plans of two silenced ventilation opening arrangements. (a) Parallel baffles and (b) lined baffles with double-door access [9].

mufflers as well as lined labyrinth ducts will be discussed in the next chapter.

6.5 Noise Reduction of a Partition Wall and Enclosure

Often, one needs to relate SPL outside (L_{p2}) with the SPL inside (L_{p1}) an existing enclosure or room with access openings and ventilation louvers. Difference of the two SPLs is called Noise Reduction, NR. Thus,

$$NR = L_{p1} - L_{p2} \qquad (6.30)$$

Prediction of NR is particularly needed in an Environmental Impact Assessment (EIA) exercise. Towards this end, some useful relationships are reproduced below.

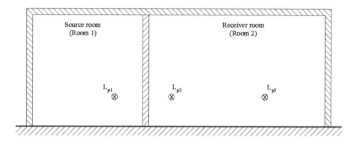

Fig. 6.7   Schematic of partition wall for illustration of noise reduction (NR).

With reference to Fig. 6.7, NR is given by [1]

$$NR = L_{p1} - L_{p2} = TL - 10\log\left(\frac{1}{4} + \frac{S_w}{R_2}\right) \text{ dB} \qquad (6.31)$$

where TL and S_w are transmission loss and area of the partition wall, respectively, and R_2 is the room constant of the receiver room (Room 2). As shown in Fig. 6.7, L_{p1} and L_{p2}, are the sound pressure levels on either side of, and close to, the partition wall.

It may be noted that in the absence of the receiver room; that is, when L_{p2} is measured just outside a wall in free field, then,

$$NR = TL + 6 \, \text{dB} \qquad (6.32)$$

Similarly, when the microphone in the receiver room is far from the partition wall, then [1]:

$$L_{p1} - L_{p3} = TL - 10\log\left(\frac{S_w}{R_2}\right) \qquad (6.33)$$

Equation (6.33) would hold not only far from the partition but also in most parts of the receiver room if $S_w/R_2 \gg 1/4$; i.e., if the receiver room is a reverberation room.

In fact, Eq. (6.33) is used to evaluate TL of the partition wall from the measured values of $L_{p1} - L_{p3}$, with R_2 being calculated from the measured value of the reverberation time of Room 2. Thus,

$$TL = NR + 10\log(S_w/R_2) \qquad (6.34)$$

Making use of energy balance it can be shown that L_{pi} and L_{po}, the sound pressure level inside and immediately outside the enclosure, respectively, are related as follows:

$$NR = L_{pi} - L_{po} = TL - C = IL \qquad (6.35)$$

where constant C is given in Table 6.3 and TL is the transmission loss of the enclosure walls. Thus, noise reduction of an acoustic enclosure is equal to its insertion loss.

To predict the average SPL on a hypothetical parallelepiped surface, say at $1\,\text{m}$ ($d = 1\,\text{m}$) around a rectangular acoustic enclosure ($l \times b \times h$) with known insertion loss IL in a room with room constant R in the frequency band of interest, we make use of the following relationship:

$$L_p = L_w - IL + 10 \log \left\{ \frac{1}{S_m} + \frac{4}{R} \right\} \qquad (6.36)$$

where, as shown in Fig. 6.8, area of the measurement surface, S_m, is given by

$$S_m = 2(l + 2d)(h + d) + 2(b + 2d)(h + d) + (l + 2d)(b + 2d) \qquad (6.37)$$

Example 6.4. A diesel generator (DG) set with sound power level of $120\,\text{dB}$ in the $500\,\text{Hz}$ octave band is located in the middle of the floor of a $20\,\text{m} \times 15\,\text{m} \times 6\,\text{m}$ industrial shed with average absorption

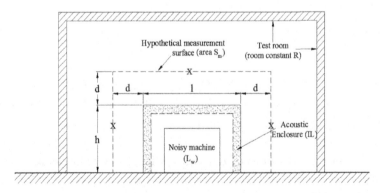

Fig. 6.8   Measurement of average SPL on a hypothetical parallelepiped surface inside a large room.

coefficient of 0.1 in the same frequency band. The DG set is provided with a $4\,m \times 3\,m \times 3\,m$ acoustic enclosure made of a 1.6 mm thick GI plate lined on the inside with 100 mm thick layer of mineral wool. Evaluate the average SPL at 1 m around the enclosure surface.

Solution. We make use of Eq. (6.36) along with Eq. (6.37) and Fig. 6.8 for the measurement surface area S_m, Eq. (6.29) along with Tables 6.2 and 6.3 for the enclosure IL, and Eq. (6.4) along with Table 6.1 for the room constant R. $L_w = 120\,dB$ (given).
Referring to Table 6.2,

$$TL \text{ of a } 1.6\,mm \text{ thick GI plate (at } 500\,Hz) = 27\,dB$$

Referring to Table 6.3,

$$\text{For average acoustic lining, } C \text{ (at } 500\,Hz) = 7\,dB$$

Therefore, $IL = TL - C = 27 - 7 = 20\,dB$
Referring to Fig. 6.8, $l = 4\,m$, $b = 3\,m$, $h = 3\,m$ and $d = 1\,m$.
Now, making use of Eq. (6.37) we get the measurement surface area, S_m,

$$S_m = 2(4+2)(3+1) + 2(3+2)(3+1) + (4+2)(3+2) = 118\,m^2$$

Area of the room surface is given by

$$S = 2(20*15) + 2*6(20+15) = 1020\,m^2$$
$$\bar{\alpha} = 0.1 \ (given)$$

Thus, making use of Eq. (6.4) the room constant (at 500 Hz) works out to be

$$R = \frac{1020 * 0.1}{1 - 0.1} = 113.3\,m^2$$

Finally, use of Eq. (6.36) yields the average value of SPL on the measurement surface

$$L_p = 120 - 20 + 10\log\left\{\frac{1}{118} + \frac{4}{113.3}\right\} = 86.4\,dB$$

6.6 Acoustics of Barriers

Acoustic barriers are often used for audio privacy between adjacent cabins in an office layout, partial protection of the road-side housing colony from the traffic noise, etc. Trees and bushes are also planted sometimes for landscaping as well as environmental noise control.

Sound diffracts around a finite barrier from all three sides. Low-frequency (or large-wavelength) waves bend around more efficiently than the high-frequency (or small-wavelength) waves. The effectiveness of barriers increases with Fresnel number, N_i, defined by [1]

$$N_i \equiv 2\delta_i/\lambda \qquad (6.38)$$

where δ_i = difference in the diffracted path and the direct path between the source and the receiver (m) and λ = wavelength, c_0/f (m).

SPL at direct distance r across the barrier in a room may be evaluated by incorporating D, the diffraction directivity factor, into Eq. (6.1). Thus, in the frequency band of interest, we can write [1]:

$$L_p(r) = L_w + 10\log\left(\frac{Q_B}{4\pi r^2} + \frac{4}{R}\right) \text{ dB} \qquad (6.39)$$

where Q_B, the barrier directivity factor is product of the location-cum-inherent directivity factor Q and the diffraction directivity factor D. Thus,

$$Q_B = Q \cdot D \qquad (6.40)$$

where

$$D = \sum_{i=1}^{3} \frac{1}{3 + 10N_i} = \sum_{i=1}^{3} \frac{\lambda}{3\lambda + 20\delta_i} \qquad (6.41)$$

If the barrier is extended to the ceiling or one of the walls, then that path is blocked and the summation in Eq. (6.41) must exclude it; the subscript "i" will no longer extend to 3. To be more precise, there would also be some contribution to D from other secondary paths because of reflections from the ground, side walls and ceiling of the room, but these can be ignored from a practical point of view (for an accuracy of $\pm 1\,\text{dB}$).

Comparing Eq. (6.39) with (6.1), insertion loss of a barrier within a room is given by

$$IL = 10\log\left(\frac{\frac{Q}{4\pi r^2} + \frac{4}{R}}{\frac{Q_B}{4\pi r^2} + \frac{4}{R}}\right)\,\text{dB} \qquad (6.42)$$

where Q_R is given by Eqs. (6.40) and (6.41).

It may be noted from Eq. (6.42) that for the barrier to be effective indoors, i.e., for a substantial value of IL, the reverberant field contribution $4/R$ must be much less than the direct field contribution $Q/(4\pi r^2)$. In other words, the surface of the room must be acoustically treated. Alternatively, r must be as small as feasible, i.e., the source as well as the receiver must be in the acoustical shadow of the barrier.

In a free field, however, the room constant R would tend to infinity, and then Eq. (6.42) would reduce to

$$IL\ (\text{free field}) = -10\log D \qquad (6.43)$$

Combination of Eqs. (6.41) and (6.43) yields

$$IL\ (\text{free field}) = -10\log\left[\lambda\left\{\frac{1}{3\lambda + 20\delta_1} + \frac{1}{3\lambda + 20\delta_2} + \frac{1}{3\lambda + 20\delta_3}\right\}\right] \qquad (6.44)$$

For a long acoustic barrier (extending several wavelengths on either side) like a highway barrier, Eq. (6.44) reduces to

$$IL\ (\text{long barrier, free field}) = -10\log\left[\frac{\lambda}{3\lambda + 20\delta_1}\right] \qquad (6.45)$$

For such a barrier, height is the only design parameter. Referring to Fig. 6.9,

$$\delta = (c + d) - (a + b) \qquad (6.46)$$

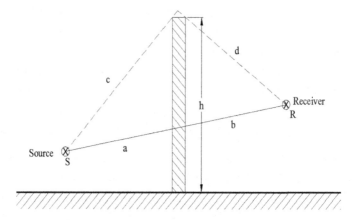

Fig. 6.9   Schematic of a long (semi-infinite) barrier of height h.

The trigonometric implications of Eq. (6.46) and Fig. 6.9 are as follows:

(a) The more the barrier height h, the more would be its insertion loss IL. As an important corollary, a highway barrier blocks tyre noise more effectively than the engine noise.
(b) Increasing the height beyond (say) 5 m is generally not a very cost-effective measure. One could explore the feasibility of building the acoustic barrier on the top of an intervening hill or earthen mound.
(c) The barrier should be located so that the source or receiver falls in its shadow, as it were. For example, a railway barrier should be located as near to the railway line or the railway colony as logistically feasible.
(d) When a highway or railway line has acoustic barriers on both sides, the barriers should be lined with an acoustically absorptive layer which can stand the elements (sun, rain, snow, etc.).

Example 6.5. It is proposed to isolate the DG set of Example 6.4 from the rest of the shed by means of a wall-to-wall 5-m high barrier as shown in Fig. 6.9, instead of a stand-alone acoustic enclosure. Evaluate the insertion loss of the barrier in the 500-Hz octave band.

Solution. We make use of Eq. (6.42) along with Eqs. (6.40) and (6.41)

Wavelength $\lambda = c_0/f = 344/550 = 0.688\,\text{m}$
As the DG set is on the floor, $Q = 2$
Assume the acoustic center of the DG set to be at a height of $1\,\text{m}$, i.e.,

$$h_s = 1\,\text{m}$$

Let the receiver be also at about the same height, so that,

$$h_R = 1\,\text{m}$$

Logistically, keeping in mind the dimensions of the DG set, let

$$a = b = 3\,\text{m}$$

Then, referring to Fig. 6.9, we have,

$$c = d = \{a^2 + (h - h_s)^2\}^{1/2}$$
$$= \{3^2 + (5 - 1)^2\}^{1/2}$$
$$= 5\,\text{m}$$

As per Eq. (6.46),

$$\delta = (5 + 5) - (3 + 3) = 4\,\text{m}$$

As the barrier stretches from wall to wall, diffraction will take place from the top only. Thus, the use of Eq. (6.41) yields,

$$D = \frac{\lambda}{3\lambda + 20\delta} = \frac{0.688}{3 * 0.688 + 20 * 4} = 0.0084$$

Distance of the receiver from the source, $r = a + b = 3 + 3 = 6\,\text{m}$
From Example 6.4, room constant (at $500\,\text{Hz}$), $R = 113.3\,\text{m}^2$
Substituting those data in Eq. (6.42) yields,

$$\text{IL} = 10\log\left[\frac{\frac{2}{4\pi*36} + \frac{4}{113.3}}{\frac{2*0.0084}{4\pi*36} + \frac{4}{113.3}}\right] = 10\log\left[\frac{0.0044 + 0.0353}{0.00004 + 0.0353}\right]$$

$$= 0.5\,\text{dB only.}$$

Thus, the barrier is practically useless. This disappointing result indicates a very important design consideration: a barrier is ineffective in a reverberant room or shed. The room, particularly its ceiling, must be acoustically lined for the wall-to-wall barrier to offer substantial insertion loss. For example, if all available area of the four walls (above a height of 2 m) and ceiling of the room were lined with an acoustical material with α (at 500 Hz) = 0.94 (see Table 6.1 for rockwool blanket of 50 mm thickness), then using Eq. (6.5),

$$\bar{\alpha} = \frac{\{2(20 + 15)(6 - 2) + 20 \times 15\} * 0.94 + 20 * 15 * 0.1}{2(20 + 15) * 6 + 2 * 20 * 15} = 0.564$$

For this value of $\bar{\alpha}$, room constant R will be,

$$R = \frac{1020 * 0.564}{1 - 0.564} = 1319\,\mathrm{m^2}$$

Then,

$$\frac{4}{R} = \frac{4}{1319} = 0.003$$

And the new value of IL is given by

$$IL = 10\log\left[\frac{0.0044 + 0.003}{0.00004 + 0.003}\right] = 3.9\,\mathrm{dB}$$

This value is typical of acoustic barriers. It is substantially higher than the value of 0.5 dB of the original (unlined) room. However, it is much lower than IL of 20 dB offered by the acoustic enclosure of Example 6.6. Therefore, for indoor use in industrial sheds, acoustic barrier is not a design option; one must use a stand-alone acoustic enclosure or an acoustic hood instead.

Other ways of increasing the IL of the barrier are shown in Fig. 6.10. As a corollary of the barrier shapes shown in this figure, any buildings, sheds, storerooms, etc., would yield high values of insertion loss. This feature is used extensively by knowledgeable architects while designing residential layouts along or near the busy highways and railway lines.

Equation (6.38) indicates that a barrier of given height will be more effective at higher frequencies than at lower frequencies. In fact,

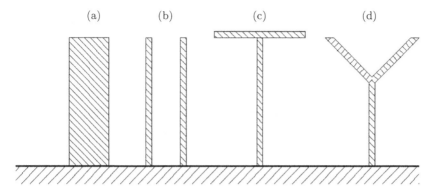

Fig. 6.10   Enhancing the insertion loss of a barrier: (a) Thick barrier, (b) compound (double) barrier, (c) barrier with a flat cap and (d) barrier with a forked top.

at very high frequencies, sound travels in a straight line like light rays, and therefore the barrier acts as a very effective reflector; there is little diffraction around the barrier.

References

[1] Irwin, J. D. and Graf, E. R., *Industrial Noise and Vibration Control*, Prentice-Hall, Englewood Cliffs, NJ, USA (1979).

[2] Sabine, W. C., *Collected Papers on Acoustics*, American Institute of Physics, New York (1993).

[3] Bies, D. A., Uses of anechoic and reverberant rooms, *Noise Control Engineering Journal*, 7, pp. 154–163 (1976).

[4] International Standards Organization, Acoustics: Measurement of sound absorption in a reverberation room, ISO 354 (1985).

[5] American Society of Testing Materials, Standard test method for sound absorption coefficient by the reverberation room method, ASTM C423-99a (1984).

[6] Bies, A. and Hansen, C. H., *Engineering Noise Control: Theory and Practice*, Fourth Edition, Spon Press, London, (2009).

[7] Harris, C. M., *Noise Control in Buildings*, McGraw Hill, New York (1994) (Appendix 3: Tables of Sound Absorption Coefficients, compiled by C.M. Harris and Ron Moulder).

[8] American National Standard 'Engineering methods for the determination of sound power levels of noise sources in a special reverberation test room' ANSI S12.33-1990 (also, ISO 3742).

[9] Crocker, M. J., *Handbook of Noise and Vibration Control*, Chapter 56, John Wiley, New York (2007).

Problems

6.1. Space-averaged sound pressure level in an $8\,\text{m} \times 5\,\text{m} \times 4\,\text{m}$ (high) reverberant room with the reverberation time of 2 seconds in the 1000 Hz octave frequency band, produced by a loudspeaker source of 10 W electrical power, is 100 dB. What is the acoustical to electrical power conversion efficiency of the loudspeaker in the frequency band?

[**Ans.: 0.346%**]

6.2. Reverberation time in the 500 Hz octave band of a $7\,\text{m} \times 5\,\text{m} \times 4\,\text{m}$ (high) room is 2 s when bare, and 1.3 s when a $3\,\text{m} \times 3\,\text{m}$ area of its floor is covered with an acoustical blanket under test. Evaluate the acoustical power absorption coefficient of the test blanket.

[**Ans.: 0.67**]

6.3. Average SPL due to a window air conditioner in a bare (unfurnished) $5\,\text{m} \times 4\,\text{m} \times 3\,\text{m}$ (high) room with $\alpha = 0.05$ is 60 dB in the 500 Hz octave band. If the entire ceiling of the room were covered with a 13 mm thick mineral tile (directly fixed to the ceiling slab), what would be the reduced value of *SPL* in the same frequency band?

[**Ans.: 53.6 dB**]

6.4. A portable genset, located in a corner of the floor of $8\,\text{m} \times 5\,\text{m} \times 3\,\text{m}$ (high) room with average sound absorption coefficient of 0.1 in the 250 Hz octave band, radiates sound power level of 86 dB in that band. Estimate SPL in the middle of the room. At what distance, will the direct sound field be equal to the diffuse field?

[**Ans.: 80 dB; 1.67 m**]

6.5. Making use of the Mass Law, evaluate *TL* for a 10 cm thick aluminum panel over all the eight major octave frequency bands (63–8000 Hz), when the panel is

(a) immersed in water,

(b) in air at 25°C.

[Ans.: Water: 0.0, 0.0, 0.1, 0.3, 1.2, 3.7, 8.0, 13.4 dB

Air: 42.2, 48.2, 56.3, 60.3, 66.3, 72.3, 78.3, 86.4 dB]

6.6. What should be the minimum thickness of the impervious steel plate of an acoustic enclosure, lined on the inside with 100 mm thick, 64 kg/m^3 density mineral wool so that the enclosure has an insertion loss of at least 25 dB at 500 Hz.

[Ans.: 2.1 mm]

6.7. An out-door compressor is generating a sound power level (SWL) of 120 dB at 500 Hz. The Government requires sound pressure level (SPL) at the property line 10 m away not to exceed 75 dB during the day and 70 dB during the night operation in an industrial area. Evaluate the minimum insertion loss for which the acoustic enclosure for the compressor should be designed in order to meet the statutory limit for 24-hour operation of the compressor.

[Ans.: 22 dB]

6.8. A railway line is provided with a sufficiently long, 5 m high acoustic barrier in order to give some acoustical protection to a parallel row of residential houses 50 m away. The barrier is located 5 m away from the centerline of the railway tracks. Assuming the source to be at a height of 1 m from the ground, what would be the noise reduction at 250 Hz for a resident at a height of

(a) 2 m (ground floor), and
(b) 6 m (first floor)?

[Ans.: (a) 13.9 dB, (b) 13.0 dB]

6.9. A 4 m high and 5 m long acoustic barrier is erected in the center of a 10 m × 8 m × 5 m (high) room with average sound power absorption coefficient of 0.2 in the 500 Hz octave band, as shown in the following figure. What is the insertion loss of the barrier in this same frequency band for a set of source (S) and receiver (R) if both are located 1.5 m from the ground and 3 m on opposite sides of the barrier? What would be the *IL* of the same barrier in free field?

[Ans.: 0.27 dB; 12.2 dB]

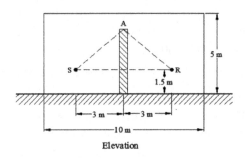

Elevation

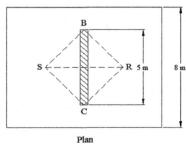

Plan

Chapter 7

Mufflers and Silencers

7.1 Introduction

The terms "muffler" and "silencer" are often used interchangeably. They are used extensively on the intake as well as exhaust systems of the reciprocating internal combustion engines, compressors, fans, blowers, gas turbines, the heating ventilation and air-conditioning (HVAC) systems, high-pressure vents and safety valves. Practically, the intake and/or exhaust (or discharge) systems of all flow machinery are fitted with mufflers. Automotive engines are invariably provided with exhaust mufflers. Therefore, the theory and practice of exhaust mufflers has history of over a 100 years.

Passive mufflers are of two types: the reactive (or reflective) type and absorptive (or dissipative) type. Reactive mufflers work on the principle of impedance mismatch. Making use of sudden changes in the area of cross-section, perforated elements, resonators, etc., the incoming or incident energy is reflected back to the source. In fact, a combination of such elements helps to reduce the acoustic load resistance faced by the source to much less than the characteristic impedance of the exhaust (or intake) pipe so that the source produces much less noise than it would produce into an anechoic load (or termination). An absorptive muffler, on the other hand, does not alter the sound produced by the source, but converts it into heat as sound propagates through its absorptive passages — acoustically absorptive linings.

Active noise control (ANC) in a duct consists in making use of a secondary source of noise and an adaptive digital control system in order to produce nearly zero impedance at the junction. Thus, an ANC system produces an acoustical short-circuit effectively muffling the primary as well as secondary sources of sound. ANC system is most effective at low frequencies (50–500 Hz) whereas passive mufflers work best in the middle and high frequencies. The reactive mufflers, however, can be specially configured for low-frequency attenuation, as will become clear later in the chapter.

Design or selection of a muffler is based on the following considerations:

(i) Adequate insertion loss so that the exhaust (or intake) noise is reduced to the level of the noise from other components of the engine (or compressor or fan, as the case may be), or as required by the environmental noise pollution limits.

(ii) Minimal (or optimal) mean pressure drop so that the source machine does not have to work against undue or excessive back pressure. (This is particularly applicable to fans or blowers which would stall under excessive back pressure.)

(iii) Size restrictions, particularly under the vehicle.

(iv) Weight restrictions.

(v) Durability, particularly in view of sharp thermal gradients in rain or on wet roads.

(vi) Cost-effectiveness is often the most important design criterion.

7.2 Electro-Acoustic Modeling

We advocate here use of the direct electro-acoustic analogies, where acoustical pressure and mass (or volume) velocity correspond to the electromotive force (or voltage) and current, respectively. Thus, acoustical impedance would correspond to the electrical impedance, and we can make use of the electrical analogous circuit for representing one-dimensional (1D) acoustical filters or mufflers as shown in Fig. 7.1.

As per Thevenin theorem, working in the frequency domain, any linear time-invariant source may be presented by an open-circuit voltage or its acoustical analogue p_s and internal impedance $Z_{s,}$ in line with the source as shown in Fig. 7.1(a). This is called the voltage

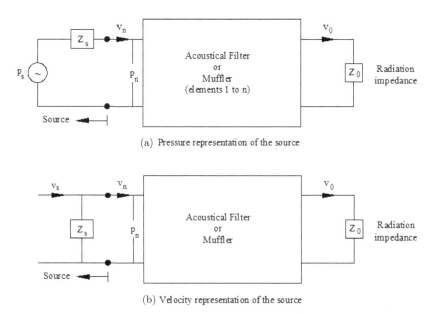

(a) Pressure representation of the source

(b) Velocity representation of the source

Fig. 7.1   Electrical analogous circuit of the muffler, radiation impedance and two equivalent representations of the source.

(or pressure) representation of the source. Alternatively, as per Norton theorem, the sources can be represented by a current, or its acoustical analogue, mass velocity v_s, with source impedance Z_s in the shunt position, as shown in Fig. 7.1(b). It can be easily shown [1] that the two representations are equivalent — they deliver the same current or velocity against a given load impedance — provided $v_s = p_s/Z_s$.

Acoustic load impedance faced by the source is given by

$$\zeta_n = p_n/v_n \qquad (7.1)$$

When $Z_s \to 0$, then Fig. 7.1(a) shows that $p_n = p_s$ for all values of the load impedance ζ_n, and the source acts as a constant pressure source. When $Z_s \to \infty$, then Fig. 7.1(b) shows that $v_n = v_s$ for all values of the load impedance ζ_n, and the source acts a constant velocity source. Thus, a zero-impedance source is a constant pressure source and an infinite-impedance source is a constant velocity source.

For use on an engine, the muffler consists of an exhaust pipe, the muffler proper and tail pipe, as shown in Fig. 7.2.

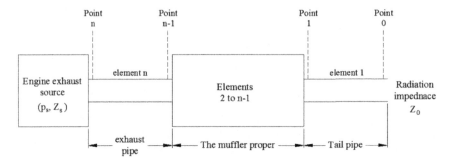

Fig. 7.2 Nomenclature of acoustical elements and points for a typical engine exhaust system.

In the absence of any muffler, the source would face the radiation impedance Z_0 instead of ζ_n. For typical applications (engines), the effect of the mean temperature and density gradients is relatively small, and the exhaust pipe diameter is generally equal to the tail pipe diameter, and therefore, $Y_n = Y_1$. Referring to Fig. 7.1(b), it can be shown that insertion loss of a given muffler is given by [1]

$$IL = 20\log\left|\frac{Z_s}{Z_s + Z_0}\frac{v_s}{v_0}\right| \qquad (7.2)$$

It can also be shown [1] that transmission loss is the limiting value of IL for anechoic source ($Z_s = Y_n$), anechoic load ($Z_0 = Y_1$), and $Y_n = Y_1(= Y_0, say)$. Symbolically,

$$TL = Lim\ IL \quad \text{as } Z_s = Z_0 \to Y_0 \qquad (7.3)$$

where Y_0 is the characteristic impedance of the exhaust pipe as well as the tail pipe (the two are assumed here to be of the same cross-section).

7.3 Transfer Matrix Modeling

The velocity ratio v_s/v_0 in Eq. (7.2) may be calculated by means of the transfer matrix method, which is ideally suited for analysis of 1D systems like acoustical filters (or mufflers), vibration isolators, electrical wave filters, etc. Making use of the matrizant approach along with the basic governing equations, namely the mass continuity equation, momentum equation, isentropicity relation, and working in the frequency domain (that is, assuming harmonic time dependence $e^{j\omega t}$ for both the state variables, p and v at all points of the muffler), one can derive transfer matrices for a variety of elements that constitute the present-day automotive mufflers [1, 2]. Some of these basic elements are shown in Figs. 7.3–7.36. In each of these figures, points u and d denote "upstream" and "downstream," respectively. Transfer matrices connect state variables p and v at point u and those at point d.

For example, the transfer matrix relationship for a rigid wall uniform area pipe or tube (a 1D waveguide) is given by [1]

$$\begin{bmatrix} p_u \\ v_u \end{bmatrix} = e^{-jMk_cl} \begin{bmatrix} \cos(k_cl) & jY_0\sin(k_cl) \\ j\sin(k_cl)/Y_0 & \cos(k_cl) \end{bmatrix} \begin{bmatrix} p_d \\ v_d \end{bmatrix} \tag{7.4}$$

where p_u and v_u are the upstream state variables (acoustic pressure and mass velocity), p_d and v_d are the downstream state variables.

Fig. 7.3   Uniform diameter, rigid wall tube/duct/pipe.

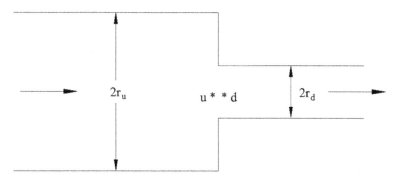

Fig. 7.4   Sudden contraction.

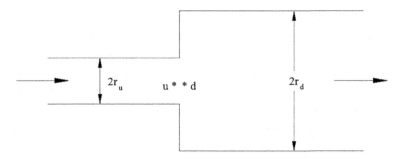

Fig. 7.5   Sudden expansion.

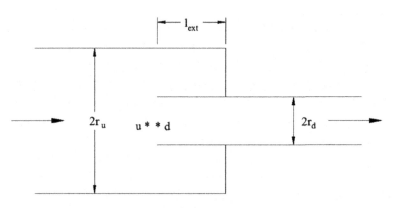

Fig. 7.6   Extended outlet.

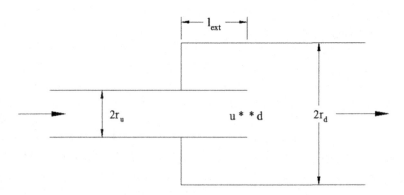

Fig. 7.7   Extended inlet.

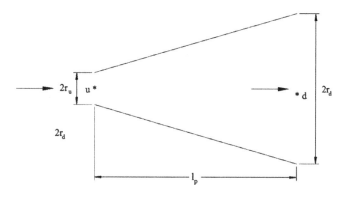

Fig. 7.8   Conical duct.

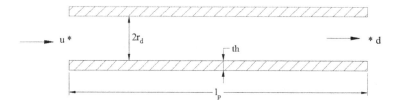

Fig. 7.9   Hose (or a uniform area tube with compliant wall).

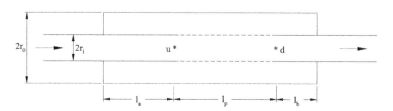

Fig. 7.10   Concentric tube resonator.

The 2×2 square matrix connecting the two state vectors is the transfer matrix of the uniform pipe with moving medium,

$k_c = \frac{k_0}{1-M^2}$ is the convective wave number,

$k_0 = \omega/c_0$ is the stationary wave number,

$Y_0 = c_0/S$ is the characteristic impedance of the pipe,

S and l are the area of cross-section and length of the pipe, respectively,

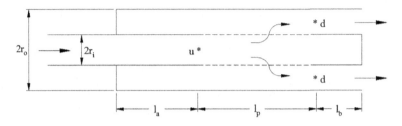

Fig. 7.11   Crossflow expansion.

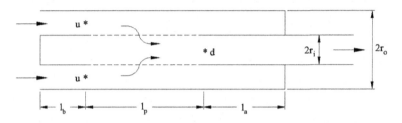

Fig. 7.12   Crossflow contraction.

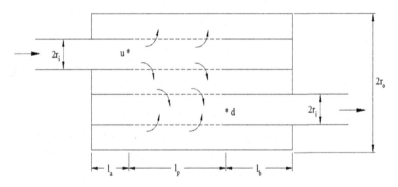

Fig. 7.13   Crossflow, 3-duct, closed-end element.

$M = U/c_0$ is the mean flow Mach number, and
U is the mean flow axial velocity averaged over the cross-section.

Typically, in the case of the intake/exhaust systems of engines, compressors, fans and HVAC systems, the mean flow Mach number M is of the order of 0.1. Therefore, $M^2 \ll 1$, $k_c \simeq k_0$, and

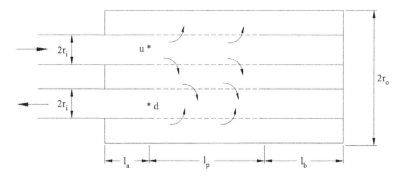

Fig. 7.14 Reverse-flow, 3-duct, closed-end element.

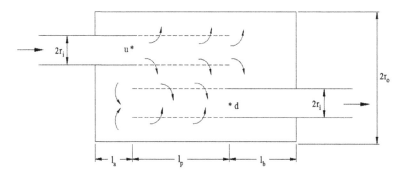

Fig. 7.15 3-Duct, open-end perforated element.

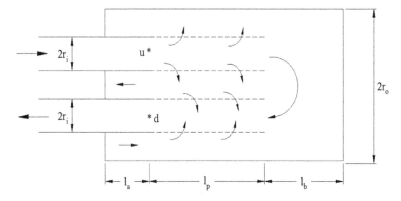

Fig. 7.16 Reverse-flow, 3-duct open-end perforated element.

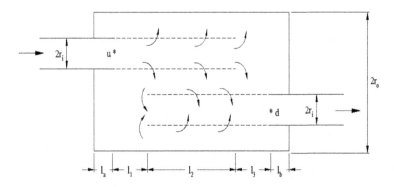

Fig. 7.17 Extended (non-overlapping) perforation crossflow, open-end element.

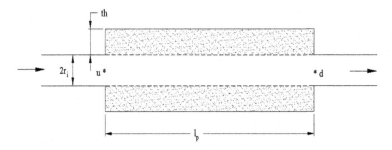

Fig. 7.18 Acoustically lined duct.

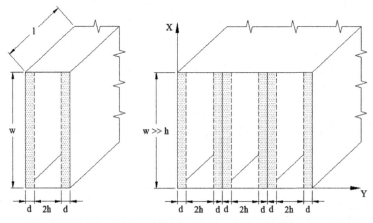

Fig. 7.19 Parallel baffle muffler and a constituent rectangular duct lined on two sides.

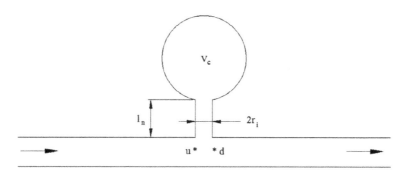

Fig. 7.20   Helmholtz resonator.

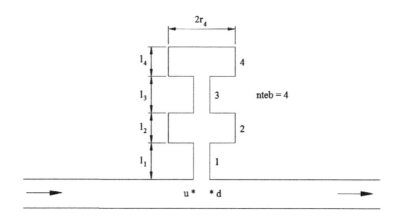

Fig. 7.21   Branch sub-system.

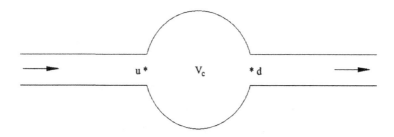

Fig. 7.22   Inline cavity.

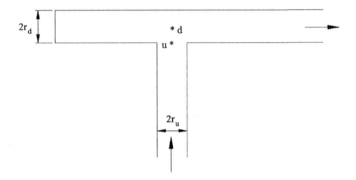

Fig. 7.23   Side inlet.

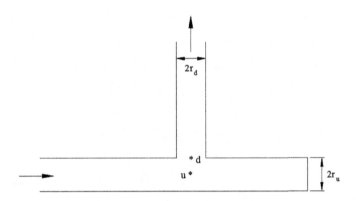

Fig. 7.24   Side outlet.

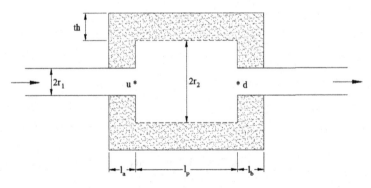

Fig. 7.25   Acoustically lined plenum chamber.

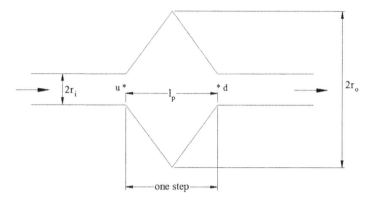

Fig. 7.26   Compliant bellows.

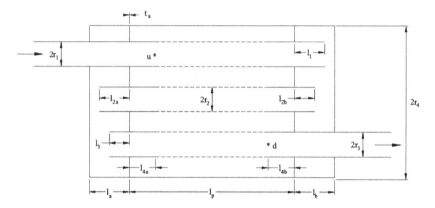

Fig. 7.27   Extended-tube 3-pass perforated element chamber.

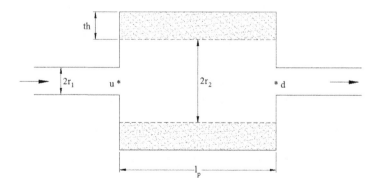

Fig. 7.28   Lined wall simple expansion chamber.

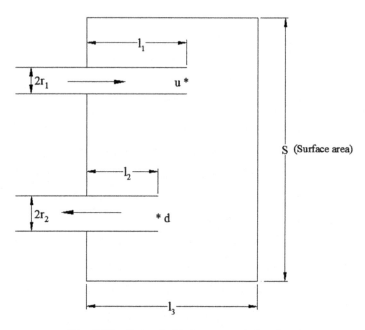

Fig. 7.29   Extended tube reversal chamber.

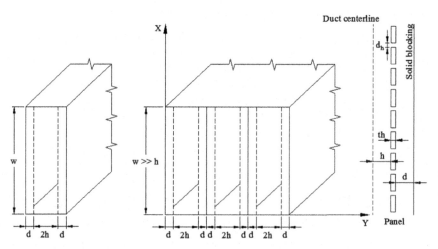

Fig. 7.30   Micro-perforated Helmholtz panel parallel baffle muffler and a constituent rectangular duct.

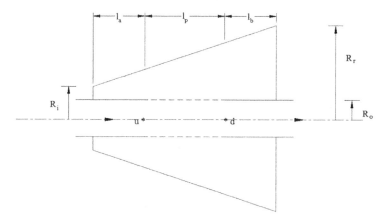

Fig. 7.31   Conical concentric tube resonator (CCTR).

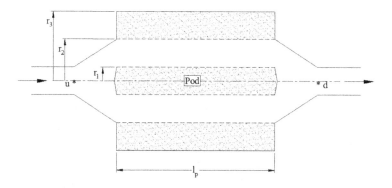

Fig. 7.32   Pod silencer.

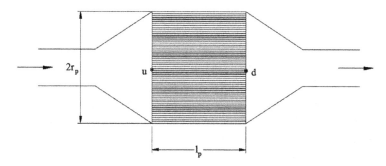

Fig. 7.33   Catalytic converter (capillary-tube monolith).

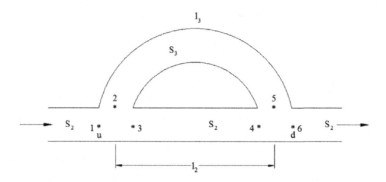

Fig. 7.34   Quincke tube.

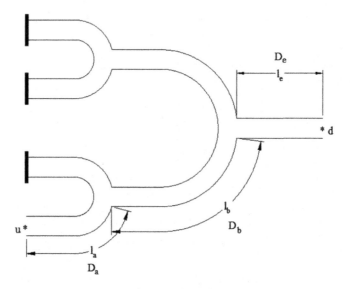

Fig. 7.35   Equal-length 4-cylinder runner manifold.

the transfer matrix in Eq. (7.4) may well be approximated by its stationary-medium counterpart:

$$\begin{bmatrix} \cos(k_0 l) & jY_o \sin(k_0 l) \\ \frac{j\sin(k_0 l)}{Y_0} & \cos(k_0 l) \end{bmatrix} \tag{7.5}$$

In fact, the convective effect of mean flow, by virtue of which the forward wave moves faster $(c_0 + U)$ and the reflected wave moves

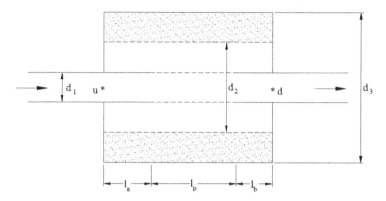

Fig. 7.36   Annular air gap lined duct.

slower $(c_0 - U)$, may be neglected in the whole of muffler analysis as a first approximation. However, the dissipative effect of mean flow at the area discontinuities and perforates, because of flow separation, plays a crucial role in the muffler performance and therefore must be incorporated appropriately.

A muffler can often be conceptualized as a cascade of some of the basic elements shown in Figs. 7.3–7.36. This is illustrated in Fig. 7.37. Comparison of the muffler of Fig. 7.37 with basic elements shown in Figs. 7.3–7.7 indicates that elements 1, 3, 5, 7 and 9 are uniform tubes, element 2 is a sudden contraction, element 8 is a sudden expansion, element 4 is an extended inlet and element 6 is an extended outlet. The overall transfer matrix of the muffler proper can then be constructed by successive multiplication of the transfer matrices of its constituent elements. Let this overall transfer matrix be denoted as

$$[T] \equiv \begin{bmatrix} T_{11} & T_{12} \\ T_{21} & T_{22} \end{bmatrix} \tag{7.6}$$

Then, it can be shown that TL is given by the expression [1]:

$$TL = 10 \log \left[\left\{ \frac{1 + M_u}{1 + M_d} \right\}^2 \frac{Y_d}{Y_u} \left| \frac{T_{11} + T_{12}/Y_d + T_{21} \cdot Y_u + T_{22} Y_u/Y_d}{2} \right|^2 \right] \tag{7.7}$$

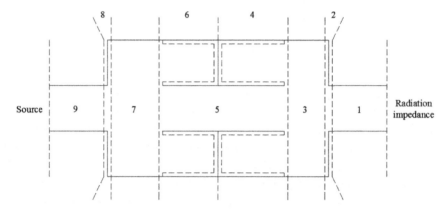

Fig. 7.37   Decomposition of a muffler into basic constituent elements.

When $Y_d = Y_u($ or, $S_d = S_u)$, then Eq. (7.7) simplifies to

$$TL = 20 \log \left| \frac{T_{11} + T_{12}/Y_0 + T_{21} \cdot Y_0 + T_{22}}{2} \right| \qquad (7.8)$$

where $Y_0 = Y_d = Y_u$. It may be recalled that subscripts u and d denote upstream and downstream, respectively.

This expression applies within the assumptions and simplifications described above and is often sufficient for synthesizing a preliminary muffler configuration [1, 3–5].

7.4 Simple Expansion Chamber (SEC)

The simplest muffler configuration is a simple expansion chamber (SEC) shown in Fig. 7.38. It was first conceptualized in the lumped-element model as a compliance (capacitance) sandwiched between two inertances (inductances), analogous to a low-pass filter (LPF) in the electrical network theory. For stationary medium, making use of the distributed-element transfer matrix of Eq. (7.5) for the chamber of length l, and combining it with the TL expression (7.7) yields the classical relationship [6].

$$TL = 10 \log \left[1 + \left\{ \frac{m - 1/m}{2} \sin(k_0 l) \right\}^2 \right] \qquad (7.9)$$

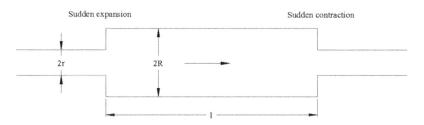

Fig. 7.38   A simple expansion chamber.

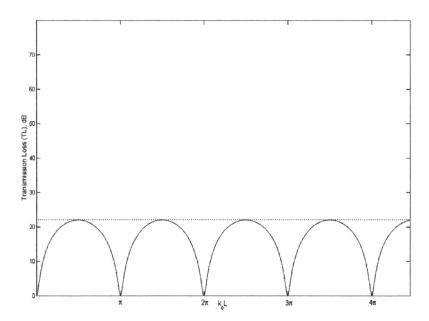

Fig. 7.39   *TL* of a simple expansion chamber.

where m is the area expansion ratio, $m = (R/r)^2$, and the product $k_0 l$ is called Helmholtz number or non-dimensional frequency.

Equation (7.9) is plotted in Fig. 7.39 for $m = 25$. The following features are noteworthy:

(i) The *TL* curve of a simple expansion chamber consists of periodic domes, with sharp troughs occurring at integral multiples of π, and peaks occurring at odd multiples of $\pi/2$.

(ii) Peak value of TL is approximately $20 \log(m/2)$ for $m \gg 1$. Thus, the larger the expansion ratio, the higher the peaks (or domes).

(iii) As $m \to 1$, $TL \to 0\,\mathrm{dB}$. This shows that the TL of a uniform pipe is zero.

(iv) Observations (ii) and (iii), when read together, indicate that in reactive mufflers, a sudden change in area of cross-section S results in a sudden change or jump in characteristic impedance $Y_0(= c_0/s)$. In fact, making use of Eq. (7.7) it can be shown [1] that in stationary medium for sudden expansion (Fig. 7.5) and sudden contraction (Fig. 7.4), both characterized by a unity transfer matrix (representing equality of pressure and mass velocity across the junction) TL is given by the common expression:

$$TL = 10 \log \left[\frac{(S_u + S_d)^2}{4 S_u S_d} \right] \tag{7.10}$$

This is the basis of the concept of "impedance mismatch" which is a fundamental principle of reactive or reflective mufflers.

Example 7.1. For a simple expansion chamber shown in Fig. 7.38, radius of the exhaust pipe and tail pipe is $20\,\mathrm{mm}$, and radius and length of the expansion chamber are 60 and $300\,\mathrm{mm}$, respectively. Find the lowest frequency range in which its TL is at least $10\,\mathrm{dB}$. Assume the medium to be air at the standard atmospheric pressure and $25°\mathrm{C}$ temperature.

Solution. $r = 20\,\mathrm{mm}$, $R = 60\,\mathrm{mm}$, $l = 300\,\mathrm{mm}$, c_0 at $25°\mathrm{C} = 346\mathrm{m/s}$, $TL \geq 10\,\mathrm{dB}$.

Area expansion ratio $m = (R/r)^2 = (60/20)^2 = 9$.

Referring to Eq. (7.9) we have,

$$10 = 10 \log \left[1 + \left\{ \frac{9 - 1/9}{2} \sin(k_0 l) \right\}^2 \right]$$

This gives $\sin(k_0 l) = 0.675$.

Thus, referring to the first or lowest frequency dome of Fig. 7.39,

$$k_0 l = 0.741 \quad \text{and} \quad \pi - 0.741 \quad \text{radians}$$

or

$$\frac{2\pi f}{346} * 0.3 = 0.741 \quad \text{and} \quad 2.4$$

or

$$f = 136 \, \text{Hz} \quad \text{and} \quad 440.5 \, \text{Hz}$$

Hence, for the specified simple expansion chamber, TL would be 10 dB or more in the frequency range 136–440.5 Hz.

7.5 Extended-Tube Expansion Chamber

Schematic of a typical extended-tube expansion chamber (ETEC) is shown in Fig. 7.40. It indicates that it consists of extended inlet tube and extended outlet tube of extension lengths l_a and l_b, respectively. An extended tube acts as a branch or shunt impedance in an electrical analogous circuit where the mass velocity (or current) gets divided. The two branch impedances in Fig. 7.40 are given by [1]

$$Z_a = -jY_a \cot(kl_a) \quad \text{and} \quad Z_b = -jY_b \cot(kl_b) \qquad (7.11)$$

where k is wave number, $k = \omega/c_0; Y_a = Y_b = c_0/S_{ann}$ is the characteristic impedance of the annulus; and S_{ann} is the area of cross-section of the annulus.

Transmission loss of the ETEC of Fig. 7.40 would tend to infinity (i.e., no sound would be transmitted downstream) if Z_a or Z_b tended

Fig. 7.40   An extended-tube expansion chamber.

to zero (analogous to a short-circuit in the electrical network). This would happen when,

$$kl_a = (2m - 1)\pi/2 \quad \text{or} \quad kl_b = (2n - 1)\pi/2 \qquad (7.12)$$

where m and n are integers $(m, n = 1, 2, 3 \ldots)$.

This interesting property can be utilized to cancel some of the troughs n the TL curve (Fig. 7.39) of the corresponding simple expansion chamber Fig. 7.38) of length l. It can be shown that [4, 7]:

if $l_a = l/2$, then troughs $1, 3, 5, 7, \ldots$ would be nullified/tuned out, and

if $l_b = l/4$, then troughs $2, 6, 10, 14, \ldots$ would be nullified/tuned out.

$$(7.13)$$

This is shown in Fig. 7.41. It may be noted that all troughs except $4, 8, 12, \ldots$ are tuned out. Moreover, there is an overall lifting of the peak value of TL. This is the basic principle of a double-tuned ETEC [4, 7].

In actual practice, however, relations (7.12) and (7.13) do not hold because they do not take into account the effect of 3D evanescent waves generated at the area discontinuities. This effect manifests itself as end corrections such that [8, 9]:

Acoustic length = geometrical (or physical) length + end correction

The end corrections of sudden expansion and sudden contraction have been calculated in the literature analytically [8] and numerically [9]. The numerical methods make use of FEM or BEM techniques. Similarly, for the extended inlet and outlet, the end correction has been evaluated numerically and experimentally [7]. Extensive parametric studies [10] have resulted in the following empirical expression for the end corrections for the extended inlet/outlet:

$$\frac{\delta}{d} = a_0 + a_1 \left(\frac{D}{d}\right) + a_2 \left(\frac{t_w}{d}\right) + a_3 \left(\frac{D}{d}\right)^2 + a_4 \left(\frac{D}{d}\frac{t_w}{d}\right) + a_5 \left(\frac{t_w}{d}\right)^2$$

$$(7.14)$$

Here, $a_0 = 0.005177$, $a_1 = 0.0909$, $a_2 = 0.537$, $a_3 = -0.008594$, $a_4 = -5.425$; d and t_w are diameter and wall thickness of the inner inlet/outlet tube; and D is the (equivalent) shell diameter.

It was noted that the end correction for a given extended length is the same whether it is used for extended inlet or for extended outlet [7].

Relations (7.13) are modified now as follows:

$$l_{g,a} = \frac{l}{2} - \delta \quad \text{and} \quad l_{g,b} = \frac{l}{4} - \delta \qquad (7.15)$$

Predictions of the 1D model match with those of the 3D model only when we add the end corrections to the geometric lengths for use in the 1D model, as shown in Fig. 7.41.

Practically, double tuning aims at tuning out the first three troughs in the *TL* curve of the corresponding simple expansion chamber. This is because the unmuffled noise of engines, compressors,

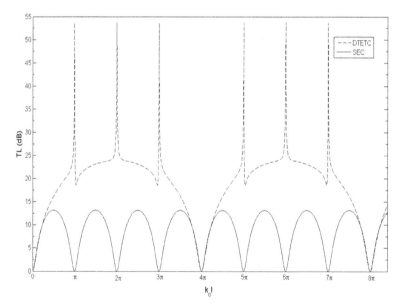

Fig. 7.41 Comparison of *TL* of a double-tuned expansion chamber with that of a simple expansion chamber.

fans, etc. predominates at lower frequencies (up to 1000 Hz) and the higher-order modes get cut on (start propagating) at higher frequencies.

Example 7.2. Re-design the simple expansion chamber of Example 7.1 as a double-tuned ETEC. Assume that the wall thickness of the inner tube is 1.25 mm. Calculate the lowest range of frequency over which TL would exceed 10 dB.

Solution. Referring to the ETEC of Fig. 7.40, we make use of Eqs. (7.14) and (7.15) to determine the required extensions $l_{g,a}$ and $l_{g,b}$ as follows:

$$d = 2r = 40\,\text{mm}, \quad D = 2R = 120\,\text{mm}, \quad D/d = 120/40 = 3,$$

$$l = 300\,\text{mm}, \quad t_w = 1.25\,\text{mm}$$

Substituting these dimensions in Eq. (7.14) yields

$$\frac{\delta}{40} = 0.005177 + 0.0909 * 3 + 0.537 \left(\frac{1.25}{40} \right) - 0.008594(3)^2$$

$$+ 0.02616 \left(\frac{3 * 1.25}{40} \right) - 5.425 \left(\frac{1.25}{40} \right)^2$$

$$= 0.2145$$

Thus, the end correction $\delta = 40 \times 0.2145 = 8.6\,\text{mm}$.

Substituting the values of chamber length l and end correction δ in Eq. (7.15) gives the required physical or geometric lengths of the extended inlet, $l_{g,a}$ and extended outlet, $l_{g,b}$:

$$l_{g,a} = \frac{300}{2} - 8.6 = 141.4\,\text{mm}$$

$$l_{g,b} = \frac{300}{4} - 8.6 = 66.4\,\text{mm}$$

Extending the frequency-range logic of Example 7.1 to Fig. 7.41 for the double-tuned extended tube expansion chamber muffler, we find that TL would exceed $10\,\mathrm{dB}$ over the range,

$$k_0 l = 0.741 \text{ to } 4\pi - 0.741 \quad \text{radians}$$

or

$$\frac{2\pi f}{346} * 0.3 = 0.741 \text{ to } 11.825 \quad \text{radians}$$

or

$$f = 136\,\mathrm{Hz} \text{ to } 2171\,\mathrm{Hz}$$

Comparing this wide range with that of the corresponding simple expansion chamber of Example 7.1 (136–440.5 Hz), reveals the tremendous advantage of double tuning, as may be noted from Fig. 7.41.

7.6 Extended Concentric Tube Resonator

Schematic of a concentric tube resonator (CTR) is shown in Fig. 7.10. When combined with the ETEC of Fig. 7.40, it results in an extended concentric tube resonator (ECTR) as shown in Fig. 7.42.

Following the control volume approach of Sullivan and Crocker [10], analysis of a CTR involves modeling of the convected plane wave propagation in the inner perforated tube and the annular cylindrical cavity coupled thorough impedance of the perforation [1, Chapter 3]. For large porosity, TL of a CTR resembles the domed structure of the TL of the corresponding simple expansion chamber. Therefore, the ECTR of Fig. 7.42 may be looked upon as an ETEC

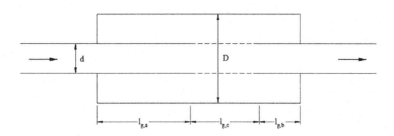

Fig. 7.42 Schematic of an ECTR.

with a perforated bridge. This point of view is particularly exciting because we can try to tune it like we did for the ETEC.

Provision of a perforate bridge between the inlet and outlet of an ETEC has advantages of little aerodynamic noise, minimal pressure drop and increased mechanical strength and durability. However, acoustic action of the resultant ECTR is very different from the corresponding double-tuned extended-tube chamber. Distributed hole-inertance of the CTR replaces the lumped inertance at the area discontinuities due to evanescent waves in the case of the corresponding extended-tube chambers.

The effective acoustic lengths are calculated precisely using 1D analysis of the ECTR. The difference of these acoustical lengths and the quarter wave resonance lengths, i.e., half and quarter of chamber lengths are termed as differential lengths. There are many variables along with temperature dependence that affect the geometrical length required to tune the ECTR, so one needs to use 1D analysis to estimate acoustical length and calculate the required physical lengths from the differential lengths and end corrections.

The differences between the two lengths (acoustical and geometric) are termed here as end corrections. These are the consequence of the inertance of perforates.

The following least-squares fit has been developed for the differential length normalized with respect to the inner-tube diameter [11],

$$\frac{\Delta}{d} = 0.6643 - 2.699\sigma + 4.522\sigma^2; \Delta = \Delta_a = \Delta_b \qquad (7.16)$$

where σ is porosity of the perforates (as a fraction). Equation (7.16) is applicable for σ ranging from 0.1 to 0.27.

Differential lengths are calculated from Eq. (7.16) for the particular porosity and inner tube diameter and these are used to estimate the initial values of acoustical lengths $l_{g,a} = \frac{l}{2} - \Delta, l_{g,b} = \frac{l}{4} - \Delta$. With the help of the 1D analysis we can increase/decrease these lengths such that the chamber length troughs are nullified effectively.

Predictions of the 1D model match with those observed experimentally, and the end corrections for this particular case are almost zero, as shown in Fig. 7.43. In particular, the first three peaks of

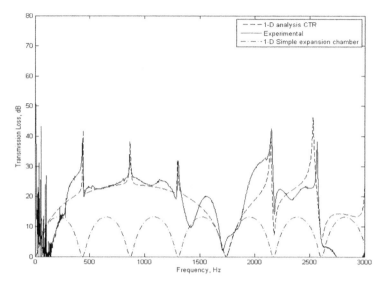

Fig. 7.43 Comparison between 1D prediction and experimental measurements for ECTR of Fig. 7.42 with 19.6% porosity [11].

the 1D curve match exactly with experimental results, as shown in Fig. 7.43.

Thus, one makes use of the 1D analysis along with precise differential lengths and end corrections to tune the ECTRs so as to lift or tune out three-fourths of all troughs that characterize the *TL* curve of the corresponding simple expansion chamber muffler. This makes the tuned ECTR a viable design option.

Example 7.3. To reduce the mean pressure drop and the aerodynamic noise generation in the ETEC muffler of Example 7.2, it is proposed to provide a perforate bridge with 19.6 % porosity so as to convert it into an ECTR, as shown in Fig. 7.42. Estimate the extended lengths $l_{g,a}$ and $l_{g,b}$.

Solution.

Given: Chamber length, $l = 0.3\,\mathrm{m}$

Inner tube diameter, $d = 2 * 0.02 = 0.04\,\mathrm{m}$

Porosity of the perforated section, $\sigma = 0.196$

The differential length Δ is given by Eq. (7.16). Thus,

$$\frac{\Delta}{0.04} = 0.6643 - 2.699 * 0.196 + 4.522(0.196)^2$$

or

$$\Delta = 0.04 * 0.31 = 0.0124\,\text{m} = 12.4\,\text{mm}$$

Now, first estimates for the extension lengths are given by

$$l_{g,a} = \frac{l}{2} - \Delta \quad \text{and} \quad l_{g,b} = \frac{l}{4} - \Delta$$

Thus,

$$I_{g,a} = \frac{0.3}{2} - 0.0124 = 0.1376\,\text{m} = 137.6\,\text{mm}$$

and

$$l_{g,b} = \frac{0.3}{4} - 0.0124 = 0.0626\,\text{m} = 62.6\,\text{mm}$$

Finally, length of the perforate (Fig. 7.42) is given by

$$l_{g,c} = l - (l_{g,a} + l_{g,b})$$
$$= 300 - (137.6 + 62.6) = 99.8\,\text{mm}$$

Starting with these estimates, we need to do a rigorous analysis of the extended-tube CTR making use of the theory outlined in Section 3.8 of Ref. [1] incorporating mean flow, in order to fine-tune $l_{g,a}$ and $l_{g,b}$ for obtaining the TL curve shown in Fig. 7.43.

7.7 Plug Muffler

Schematic of a typical plug muffler is shown in Fig. 7.44. As the flow pattern indicates, the plug forces the mean flow to move out through the perforation into the annulus and then move back through the perforation to the inner tube. Thus, a plug muffler is a combination of a cross-flow expansion element (Fig. 7.11) and a cross-flow

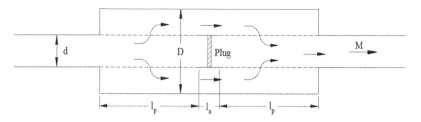

Fig. 7.44 Schematic of a plug muffler.

contraction element (Fig. 7.12) with a small uniform annular tube (Fig. 7.3) in between. Cross-flow elements act as in-line lumped flow-acoustic resistance. Therefore, their acoustic performance (TL), as well as mean pressure drop increase as the open area ratio (OAR) decreases and the mean flow Mach number, increases while remaining incompressible.

This important parameter is defined as

$$\text{OAR} \equiv \frac{\text{flow area of the perforated section}}{\text{cross-sectional area of the perforated tube}}$$

$$= \frac{\pi d l_p \sigma}{\frac{\pi}{4} d^2} = \frac{4 \sigma l_p}{d} \tag{7.17}$$

where l_p and σ are length and porosity of the perforated section (generally the same on both sides of the plug — see Fig. 7.44).

Figure 7.45 shows TL spectra of the plug muffler of Fig. 7.44 with OAR as a parameter. For $l_p = 20\,\text{cm}$, $d = 4\,\text{cm}$, as per Eq. (7.17), porosity works out to be $\sigma = 0.05\,\text{OAR}$. It is obvious from Fig. 7.45 that TL increases remarkably as OAR decreases. But then, as we shall see later in the section on back pressure, mean pressure drop increases drastically as OAR decreases. This calls for a compromise between back pressure and TL.

Mean flow Mach number, M, in the perforated tube also has a similar effect as may be observed from Fig. 7.46. Incidentally, the stagnation pressure drops increases in proportion to square of the mean flow velocity or Mach number.

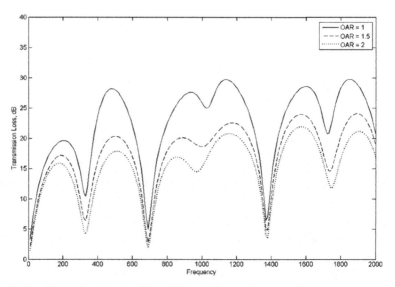

Fig. 7.45 *TL* of the plug muffler of Fig. 7.44 as a function of the open-area ratio, OAR. ($d = 4\,\text{cm}$, $D = 12\,\text{cm}$, $l_p = 20\,\text{cm}$, $l_a = 10\,\text{cm}$, $M = 0.15$.)

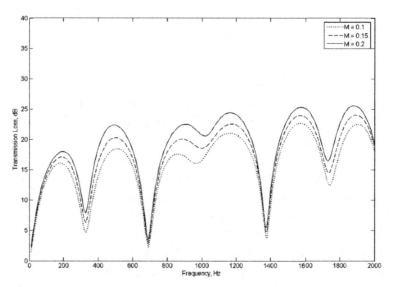

Fig. 7.46 Effect of the mean flow Mach number M on TL of a plug muffler. ($d = 4\,\text{cm}$, $D = 12\,\text{cm}$, $l_p = 20\,\text{cm}$, $l_a = 10\,\text{cm}$, OAR $= 0.15$.)

7.8 Multiply Connected Muffler

Many of the present-day automotive mufflers are elliptical in cross-section and make use of multiply connected elements. One such configuration would consist of two such end chambers connected by a uniform pipe, thereby making use of double flow reversal and inducing maximum acoustic interference.

Such a system is shown in Fig. 7.47, wherein the end-chambers (numbered 1 and 3) and connecting pipes which also act as pass tubes, are clearly shown. The lengths L_c of the connecting pipes are much greater than the length of the elliptical end chambers L_a and L_b. Rather than the time-consuming FEM process, which involves geometry creation, fine meshing (requiring a lot of computer memory, especially at higher frequencies), and solving linear systems with matrix inversion routines, a simple 1D model has been developed [13, 14]. This 1D transverse plane wave approach can be used to obtain transfer matrix which is needed to cascade it with the preceding and succeeding elements constituting a complex muffler. The direction of the transverse plane wave is taken along the major-axis of the ellipse with the cavities above the inlet and beneath the outlet modeled as variable area resonators. The impedance of such a resonator is found using a semi-numeric technique called the matrizant approach [13]. Recently, this semi-analytical method has been replaced with an analytical method, where Frobenius solution of the differential equation governing the transverse plane wave propagation is obtained [14].

It may be noted from Fig. 7.47 that the flow has more than one parallel paths. Therefore, one needs to do a flow resistance network analysis and then evaluate the resistances of the perforated portions as well as the sudden expansions and contractions for use in the acoustic analysis [15].

In a multiply connected muffler, waves in different perforated pipes have two sets of interactions with substantial frequency-dependent phase difference in between. This phenomenon is similar to that of Quincke tube (see Fig. 7.34). This results in acoustic interference which helps to raise the TL curve over a wider frequency range.

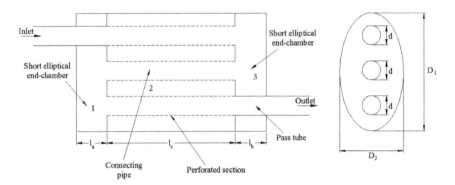

Fig. 7.47   Double-reversal end-chamber muffler.

Figure 7.47 shows only one of the many types of multiply connected mufflers. Some others are discussed by Panigrahi and Munjal [16].

7.9 Absorptive Ducts and Mufflers

An acoustically lined duct converts or dissipates acoustic energy into heat as a progressive wave moves along the duct, forward or backward. The lining can be a bulk reacting lining or locally reacting lining, depending on whether or not wave propagates through the body of the absorptive material. Common absorptive (or dissipative) materials are glass wool, mineral wool, ceramic wool, polyurethane (PU) foam, polyamide foam, etc. All these materials are highly porous with volume porosity of more than 95%. An acoustical material must have open (interconnected) pores unlike a thermal foam which has closed pores.

The most important physical parameter of absorptive material is flow resistivity. Referring to the schematic of a test setup shown in Fig. 7.48, flow resistivity, E, is given by

$$E = \frac{\Delta p}{Ud}(\text{Pa.s/m}^2) \tag{7.18}$$

where $\Delta p = \rho_w g \Delta h$ is the pressure drop across the sample of thickness d, and U is the mean flow velocity in the test duct averaged over the cross-section.

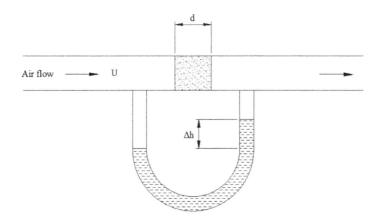

Fig. 7.48 Schematic of a setup for measuring flow resistivity.

Typically, flow resistivity E is of the order of 5000–40,000 Pa.s/m^2. Generally, for typical thickness, d(50 mm or 100 mm), the higher the value of E, the more will be value of TL of a lined duct. E increases with the mass density and fibre diameter (the smaller the fibre diameter, the more will be the value of E). An appropriate parameter is the non-dimensional acoustic flow resistance, R defined as

$$R = \frac{Ed}{\rho_0 c_0} \qquad (7.19)$$

Typically, R is of the order of 1 to 8.

The most important design parameter is h, defined as

$$h \equiv \frac{\text{cross-section of the flow passage}}{\text{wetted (or lined) perimeter}} \qquad (7.20)$$

Referring to Fig. 7.49, it may be noted that for a rectangular duct lined on two sides with an absorptive layer of thickness d, $h = b.2h/2b$. Therefore, the flow passage height is denoted by "$2h$." In other words, h is half the flow passage height.

Applying Eq. (7.20) to a circular duct lined all around with flow passage diameter D, yields,

$$h = \frac{\pi/4D^2}{\pi D} = \frac{D}{4} \qquad (7.21)$$

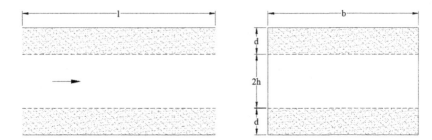

Fig. 7.49   Schematic of a rectangular duct lined on two sides.

Similar relationship would hold for a square duct lined on all four sides; h would be one-fourth of the clear passage height. The most common practice in the heating, ventilation and air-conditioning (HVAC) systems is to make use of a rectangular duct lined on two sides, as shown in Fig. 7.49.

Transmission loss of a lined duct of length l is given by

$$TL = TL_h \cdot l/h \qquad (7.22a)$$

where TL_h is the specific TL of a lined duct of length equal to h. This relationship is very helpful in as much as it indicates that TL of a duct is directly proportional to its length l and inversely proportional to h, defined by Eq. (7.20).

Equations (7.20) and (7.22a) collectively indicate that TL of a square duct lined on all four sides would be nearly double that of the duct lined on two opposite sides only.

Sometimes a designer or consultant needs to do some quick hand calculations of the effectiveness (in terms of transmission loss) of a lined duct. The datum available is $\bar{\alpha}$, the absorption coefficient of the material of the lining, defined as the fraction of the normally incident plane-wave energy absorbed by the given thickness of the lining, backed by a rigid wall. The value of the absorption coefficient $\bar{\alpha}$, supplied by the manufacturer, is an average value over a certain frequency range. There are a number of empirical formulae for quick hand calculations. One popular example is Piening's empirical formula [1], according to which,

$$TL \approx 1.5 \frac{P}{S} \bar{\alpha} l (\text{dB}) \qquad (7.22b)$$

where $\bar{\alpha}$ is the absorption coefficient of the material, P is the lined perimeter, and S is the free-flow area of the cross section.

Thus, for a circular duct of radius r_0, or a square duct with each side $2r_0$ long, lined all over the periphery,

$$TL \approx 3\bar{\alpha}(l/r_0) \approx 3\frac{l}{d_0} \qquad (7.22c)$$

Formula (7.22c) is indeed very useful for a quick estimate of the effectiveness of an acoustically lined duct. For example, it indicates that if a material with $\bar{\alpha} = 0.5$ were used to line a circular or square duct, it would yield a 3 dB attenuation across a length equal to one diameter or side length.

In order to reduce h (for increased TL from a lined duct of given length l), we make use of a parallel baffle muffler shown in Figs. 7.19 and 7.30. These figures show the cross-section or end view of a 3-pass or 3-louvre parallel baffle muffler.

It may be noted that the intermediate layer or baffle of thickness $2d$ would be servicing both the passages: the left half of it associated with the left passage and the right half with the right passage. Thus, a 3-louvre parallel baffle muffler is equivalent to three identical parallel rectangular ducts of the type shown in Fig. 7.49. Therefore, the TL-curves of Fig. 7.50 apply to a lined rectangular duct as well as a parallel baffle muffler.

Specific TL parameter in Eq. (7.22a) may be read from Fig. 7.50, drawn on the lines of Ref. [4], where it is plotted against the non-dimensional frequency parameter, η, defined as

$$\eta = 2hf/c_0 = 2h/\lambda \qquad (7.23)$$

Parameter R in Fig. 7.50 is given by

$$R = \frac{Ed}{\rho_0 c_0} \qquad (7.24)$$

where E is flow resistivity of the absorptive filling.

Although the specific TL curves shown in Fig. 7.50 are applicable for $R = 5$, yet they can be used for other values of R (say, $R = 1, 2, 10$) without much error. Therefore, the curves of Fig. 7.50 can be used practically for most commercial absorptive materials

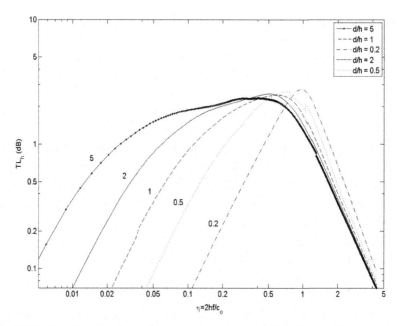

Fig. 7.50   Specific TL of a rectangular duct, or parallel baffle muffler, as a function of the non-dimensional frequency parameter for $R = 5$, illustrating the effect of d/h [4].

made out of fibres, like glass wool, mineral wool, ceramic wool, foam, etc.

The following important observations may be made from Fig. 7.50:

1. For the desired TL at lower frequency (higher wavelength), we should use wider flow passages (larger h), and vice versa.
2. For narrow band TL requirement, we should select h such that,

$$\frac{2h}{\lambda} \simeq 0.5 \quad \text{or} \quad h = \frac{\lambda}{4}$$

where λ is the wavelength corresponding the peak frequency of the unmuffled SPL spectrum.
3. For absorbing a predominantly low-frequency sound, we should use thicker baffles $(d > h)$. It is common to use $d = 2h$ for the low-frequency sound attenuation.

4. If the unmuffled A-weighted SPL spans over several octaves, or over two decades, such that η ranges from 0.02 to 2, then we may have to use $d = 5h$ $(d/h = 5)$.

5. The open area fraction is given by

$$OAF = \frac{2h}{2h + 2d} = \frac{1}{1 + d/h} \tag{7.25}$$

Thus $d/h = 1, 2$ and 5 would correspond to OAF $= 50\%, 33.3\%$ and 16.6 %, respectively.

6. Width of a square cross-section parallel baffle muffler with n_p baffles (or, for that matter, n_p passages or louvres) is given by $W = 2(h + d)n_p$ in which $2hn_p$ represents the total flow passage width. Choice of a higher value of $d(d > h)$, within a fixed value of W, would decrease h and/or n_p, and therefore carries the penalty of increased pressure drop or a larger parallel baffle muffler. Increased back pressure on the air handling fan (or blower) would:

(a) reduce the air flow,
(b) call for a more powerful motor,
(c) increase its power consumption,
(d) increase the casing noise of the fan, and
(e) increase the aerodynamic noise generated in the louvers.

A larger parallel baffle muffler would:

(a) need more space, and
(b) be heavier and costlier.

7. TL of an absorptive duct or parallel baffle muffler being proportional to its length presents an easy and sure way of increasing TL of an absorptive duct.

Observations 1 to 7 above indicate that optimum design of a parallel baffle muffler involves a compromise between acoustic attenuation, back pressure, size and cost [1, 4, 17].

Figure 7.50 applies to stationary medium. The convective effect of mean flow is to decrease TL marginally for the forward progressive wave, and increase it marginally for the rearward progressive wave (moving against the flow). This may, however, be neglected for a first design.

Example 7.4. A tube-axial fan, operating in a tube of diameter 0.6 m is producing sound power level in different octave frequency bands as follows:

Frequency (Hz)	63	125	250	500	1000	2000	4000	8000
L_w (dB)		100	106	107	106	104	103	97	95

Calculate the space-averaged, A-weighted sound pressure level at 3 m from the radiation end. If 4 m length of the 20 m long tube is lined on the inside with 100 mm thick blanket of highly porous PU foam, calculate the overall transmission loss of the lining, and the attenuated L_{PA} at the same measurement point outside the tube.

Solution. Making use of Table 1.2, the A-weighted sound power levels are tabulated below (see rows (b) and (c)).

For free field radiation, use of Eq. (6.6) gives the space-averaged, A-weighted SPL at 3 m:

$$L_{pA}(3\,\text{m}) = L_{WA} - 10\log(4\pi * 3^2)$$

$$= L_{WA} - 20.5$$

This is reflected in row (d) of the following table.

With 100 mm (=0.1 m) lining inside the 0.6 m diameter tube, clear diameter is 0.4 m. Thus, the lining thickness $d = 0.1$ m, and making use of Eq. (7.20), $h = 0.4/4 = 0.1$ m. Therefore, $d/h = 1.0, l/h = 4/0.1 = 40$, and the non-dimensional frequency parameter η in Fig. 7.50 is given by

$$\eta = \frac{2hf}{c_0} = \frac{2*0.1}{346}f = 0.000578\,f$$

Octave band center frequency (Hz)	63	125	250	500	1000	2000	4000	8000	Total
(a) L_W (dB)	100	106	107	106	104	103	97	95	112.9
(b) A-weighting correction (dB)	−26.2	−16.1	−8.6	−3.2	0.0	1.2	1.0	−1.1	—
(c) $L_{WA} = (a) + (b)$ (dB)	73.8	89.9	98.4	102.8	104.0	104.2	98.0	93.9	109.4
(d) L_{pA} (3 m) = $(c) - 20.5$ (dB)	53.3	69.4	77.9	82.3	83.5	83.7	77.5	73.4	88.9
(e) η in Fig. 7.50 = $0.000578\,f$	0.04	0.07	0.14	0.29	0.58	1.16	2.31	4.62	—
(f) TL from Fig. 7.50 for $d/h = 1$ (dB)	0.3	1.0	1.3	2.2	2.5	1.5	0.3	0.07	—
(g) $TL = TL_h \times l/h =$ $(f) \times 40$ (dB)	12	40	52	88	100	60	12	2.8	—
(h) Attenuated L_{pA} (3 m) = $(d) - (g)$ (dB)	41.3	29.4	27.9	—	—	23.7	67.5	70.6	71.8

Thus, total attenuated L_{pA} (3 m) = 71.8 dB (from row (h)), and overall transmission loss, $TL = 88.9 - 71.8 = 17.1$ dB (from rows (d) and (h)).

It is worth noting that for the data of Example 7.4 above (that is, the lined tube length, $l = 4$ m, and clear diameter $d_0 = 0.4$ m), the grossly approximate relationship of Eq. (7.22c) suggests an overall TL of $3 * 4/0.4 = 30$ dB, which is much more than the value of 17.1 dB, the end result of Example 7.4. However, it is very important that one selects h and d keeping in mind the unmuffled A-weighted SPL spectrum and Fig. 7.50. Often, additional considerations of the stagnation pressure drop, space availability and cost-effectiveness necessitate several iterations for the choice of l, h, d and the number of passes (or louvres, or parallel baffles).

7.10 Combination Mufflers

It may be observed from Fig. 7.50 that absorptive ducts and mufflers have the advantage of wide band attenuation over reactive mufflers, the TL curve of which may have sharp domes and troughs. On the other hand, the acoustic performance of absorptive ducts and mufflers is poor at low frequencies. Therefore, in order to raise the entire TL curve, particularly the troughs, it is desirable to make use of a

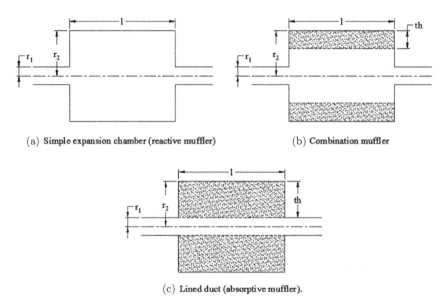

(a) Simple expansion chamber (reactive muffler)

(b) Combination muffler

(c) Lined duct (absorptive muffler).

Fig. 7.51 Schematics of muffler configurations illustrating the concept of a combination muffler ($r_2 = 3r_1$).

muffler combining the two types of mufflers [18]. A simple configuration of a combination muffler is shown in Fig. 7.51(b) and the corresponding TL spectrum is shown in Fig. 7.52, where it is compared with TL of the limiting cases of a simple expansion chamber (reactive muffler) shown in Fig. 7.51(a) and a lined duct (absorptive muffler) shown in Fig. 7.51(c). Thickness of the absorptive lining has been used as a parameter in Fig. 7.52.

7.11 Acoustic Source Characteristics of I.C. Engines

As per Thevenin theorem, analogous to electrical filter, the acoustic filter or muffler requires prior knowledge of the load-independent source characteristics p_s and Z_s, corresponding to the open circuit voltage and internal impedance of an electrical source.

Prasad and Crocker [19], based on their direct measurements of source impedance of a multi-cylinder inline CI engine, proposed the anechoic source approximation: $Z_s = Y_0$. Callow and Peat [20] came

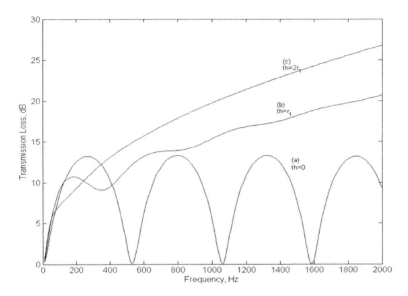

Fig. 7.52 Performance of the muffler configurations of Fig. 7.51.

out with a relatively more realistic expression:

$$Z_s(\text{exhaust}) = Y_0(0.707 - j0.707) \qquad (7.26a)$$

where Y_0 is the characteristic impedance of the exhaust pipe, c_0/S. Here, S is the area of cross-section of the exhaust pipe, and p_0 and c_0 are density and the sound speed of the exhaust gases, respectively.

Hota and Munjal [21] extended the work of Fairbrother *et al.* [22, 23] to formulate the source characteristics of a compression-ignition (CI) engine as functions of the engine's physical and thermodynamic parameters and incorporated them as empirical formulas into the scheme to predict the un-muffled noise using a multi-load method. Again, inspired by the work of Knutsson and Boden [24], the investigation of Ref. [24] was extended to the intake source characterization of C.I. engines by Hota and Munjal [25].

It has been found that internal impedance of the engine intake system is given by the empirical expression [25]

$$Z_s(\text{intake}) = Y_o(0.1 - j0.1) \qquad (7.26b)$$

Finally, Ref. [26] offers empirical expressions for the source strength level (SSL) in decibels of SI engines, for the intake as well as the exhaust system.

A pre-requisite for this investigation is to have realistic values of the pressure-time history. These have been computed here making use of the commercial software AVL-BOOST [27] for different acoustic loads. This finite-volume CFD model is used in conjunction with the 2-load method to evaluate the source characteristics at a point in the exhaust pipe just downstream of the exhaust manifold.

As per the electrical analogous circuit of the un-muffled system depicted in Fig. 7.53, for two different acoustic loads (impedances) Z_{L1} and Z_{L2}, one can write:

$$p_s Z_{L1} - p_1 Z_s = p_1 Z_{L1} \qquad (7.27)$$

and

$$p_s Z_{L2} - p_2 Z_s = p_2 Z_{L2} \qquad (7.28)$$

These two equations may be solved simultaneously to obtain:

$$p_s = p_1 p_2 \frac{Z_{L1} - Z_{L2}}{p_2 Z_{L1} - p_1 Z_{L2}} \qquad (7.29)$$

$$Z_s = Z_{L1} Z_{L2} \frac{p_1 - p_2}{p_2 Z_{L1} - p_1 Z_{L2}} \qquad (7.30)$$

It may be observed from Fig. 7.54 that for the CI engines if a least square fit is done on the SSL spectrum at different frequencies or

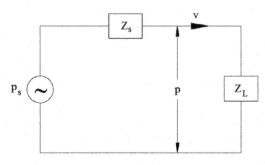

Fig. 7.53   Electrical analogous circuit for an un-muffled system [25].

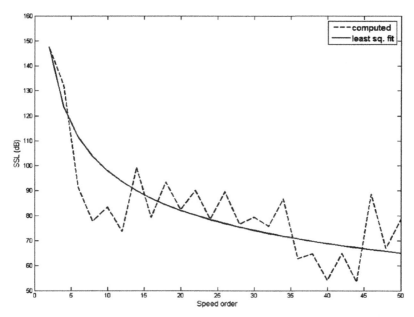

Fig. 7.54 Values of the intake SSL as a function of speed order for a turbocharged engine [25].

speed orders, the curve goes down more or less exponentially [21, 25]. Hence the generalized formula for the SSL can be defined as

$$\text{SSL} = A * \left(\frac{\text{speed order}}{N_{cyl}/2} \right)^{B} \text{dB} \tag{7.31}$$

where N_{cyl} is the number of cylinders in the 4-stroke cycle engine. $N_{cyl}/2$ represents the speed order of the firing frequency of a 4-stroke cycle engine, and constant A represents SSL at the firing frequency.

Speed order, n of frequency f_n is defined as

$$n = \frac{f_n}{RPM/60} \tag{7.32}$$

The firing frequency of a multi-cylinder engine is given by

$$\text{firing frequency} = \frac{RPM}{60} \times \frac{2}{N_{st}} \times N_{cyl} \tag{7.33}$$

As there is one firing in two revolutions of a 4-stroke ($N_{st} = 4$) cycle engine, the speed order of the firing frequency of a 4-stroke cycle engine becomes $N_{cyl}/2$.

This kind of least square fit has been done to discount sharp peaks and troughs because computations have been made by assuming that speed of the engine remains absolutely constant. But in reality, there may be around one to five percent variation in speed because the pressure-crank angle diagrams of successive cycles would never be identical.

Acoustic parametric study has been conducted for the following parameters, varying one at a time, keeping other parameters constant at their default (underlined) values:

Turbocharged diesel engines

Air fuel ratio, AFR = 18.0, **23.7**, 29.2, 38.0

Engine speed in RPM = 1000, 1300, 1600, 2100, 2400, 3000, 3500, **4000**, 4500

Engine capacity (displacement), V (in liters) = 1.0, 1.5, 2.0, **2.5**, 3.0, 4.0

Number of cylinders, N_{cyl} = 1, 2, 3, **4**, 6

So, the default turbocharged engine is: four cylinders, 2.5 L, running at 4000 rpm, with the air-fuel ratio 23.7.

Naturally aspirated diesel engines

Air fuel ratio, AFR = 14.5, **17.0**, 29.0, 39.6

Values of RPM, V and N_{cyl} are the same as for the turbocharged engine above.

A and B of SSL for Exhaust System of C.I. Engine [21]

Turbocharged diesel engines:

$$A = 171.4 * (1 - 0.0019\,AFR)(1 + 0.102\,NS - 0.016\,NS^2)$$
$$\times (1 - 0.0023\,V)(1 - 0.021\,N_{cyl})$$
$$B = -0.093 * (1 + 0.016\,AFR)(1 + 0.31\,NS - 0.076\,NS^2)$$
$$\times (1 - 0.03\,V)(1 + 0.026\,N_{cyl}) \qquad \text{(7.34a, b)}$$

Naturally aspirated diesel engines:

$$A = 167 \times (1 - 0.0015\,AFR)(1 + 0.125\,NS - 0.021\,NS^2)$$
$$\times (1 + 0.0018\,V)(1 - 0.0233\,N_{cyl})$$
$$B = -0.13 * (1 + 0.0123\,AFR)(1 + 0.19\,NS - 0.053\,NS^2)$$
$$\times (1 - 0.007\,V)(1 - 0.026\,N_{cyl}) \qquad (7.34\text{c, d})$$

where NS = engine speed in $RPM/1000$.

A and B of SSL for Intake System of CI Engine [25]
Turbocharged diesel engines:

$$A = 214 * (1 + 0.0018\,AFR)(1 - 0.08\,NS + 0.01\,NS^2)$$
$$\times (1 - 0.00021\,V)(1 - 0.05\,N_{cyl})$$
$$B = -0.283 * (1 - 0.0033\,AFR)(1 - 0.039\,NS)$$
$$\times (1 - 0.0173\,V)(1 + 0.033\,N_{cyl}) \qquad (7.35\text{a, b})$$

Naturally aspirated diesel engines:

$$A = 191.8 * (1 + 0.00075\,AFR)(1 - 0.06\,NS + 0.007\,NS^2)$$
$$\times (1 - 0.001\,V)(1 - 0.028\,N_{cyl})$$
$$B = -0.166 * (1 - 0.0015\,AFR)(1 - 0.014\,NS)$$
$$\times (1 - 0.0064V)(1 + 0.109\,N_{cyl}) \qquad (7.35\text{c, d})$$

where NS = engine speed in $RPM/1000$.

Naturally aspirated SI engines [26]
An acoustic parametric study has been conducted for the following basic parameters, varying one at a time, keeping other parameters constant at their default (underlined) values for both intake and exhaust.
Air fuel ratio, AFR = 12.0, **17.0**, 20.0, 24.0, 28.0
Engine speed in RPM = 1000, 1500, 2000, 2500, 3000, 3500, 4000, 4500, 5000, **5500**, 6000, 6500, 7000

Engine swept volume or displacement volume, V (in liters)= 0.5, 1.0, 1.5, **2.0**, 2.5, 3.0,

Number of cylinders,$N_{cyl} = 1, 2, 3, \underline{4}, 6$

Here, default parameters of the naturally aspirated SI engine are: four cylinders, 2.0 L, 5500 rpm, and the air-fuel ratio of 17. The resultant values of A and B of SSL are as follows:

Exhaust system:

$$A = 198.7 \times (1 - 0.0015\,AFR)(1 + 0.0445\,NS - 0.00765\,NS^2)$$
$$\times (1 - 0.0021\,V)(1 - 0.0374\,N_{cyl}) \tag{7.36a}$$
$$B = -0.053 * (1 + 0.008\,AFR)(1 + 0.246\,NS - 0.0314\,NS^2)$$
$$\times (1 - 0.028\,V)(1 + 0.0287\,N_{cyl}) \tag{7.36b}$$

Intake system:

$$A = 176.36 * (1 - 0.00022\,AFR)(1 + 0.0517\,NS - 0.00866\,NS^2)$$
$$\times (1 - 0.00184\,V)(1 - 0.03336\,N_{cyl}) \tag{7.37a}$$
$$B = -0.1083 * (1 + 0.0021\,AFR)(1 + 0.256\,NS - 0.032\,NS^2)$$
$$\times (1 + 0.0075\,V)(1 + 0.122\,N_{cyl}) \tag{7.37b}$$

where NS = engine speed in $RPM/1000$.

The resultant source characteristics are used with the transfer matrix-based muffler program [28] to predict the exhaust sound pressure level of a naturally aspirated 4-stroke petrol or gasoline engine. Thus, the designer will be able to compute the exhaust sound pressure level with reasonable accuracy and thereby synthesize the required muffler configuration of a spark ignition (or gasoline) engine as well as the compression ignition (or diesel) engine.

7.12 Designing for Adequate Insertion Loss

As indicated before in Section 7.2, unlike transmission loss TL, insertion loss IL depends on the source impedance and radiation impedance as well as the overall transfer matrix of the muffler proper.

While *TL* is appropriate for study of the inherent role of each element or a set of elements constituting the muffler proper, it is *IL* that is of interest to the user as a true measure of noise reduction due to a muffler. *IL* is a function of source impedance, radiation impedance, exhaust pipe, tail pipe as well as the muffler proper — see Fig. 7.2.

The *IL* spectrum for a typical muffler is quite similar to the corresponding *TL* spectrum as shown in Fig. 7.55 for a simple expansion chamber muffler of Fig. 7.38. Yet there are extra dips or troughs because of the exhaust pipe and tail pipe resonances. The most important and disturbing feature of the *IL* spectrum is a prominent dip at low frequencies (typically, lower than 100 Hz). Here, *IL* is prominently negative indicating that at this frequency and its neighborhood, the exhaust *SPL* would be higher with muffler than that without it. In other words, at this frequency, the muffler would augment noise instead of muffling it! This low-frequency dip (or dips) occurs for any source impedance (see, for example, Fig. 8.7 in Ref. [1]).

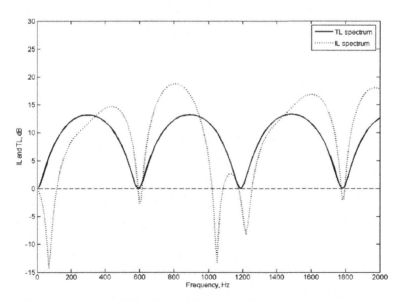

Fig. 7.55 Comparison of the *IL* and *TL* spectra for the simple expansion chamber muffler configuration of Fig. 7.38. ($c_0 = 592\,\text{m/s}, r = 2\,\text{cm}, R = 6\,\text{cm}, l_e = 0.8\,\text{m}, l = 0.5\,\text{m}, l_t = 0.3\,\text{m}, M = 0$.)

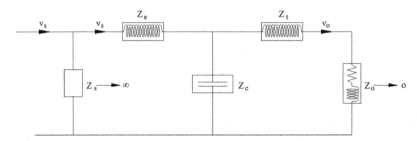

Fig. 7.56   Low-frequency (lumped-element) representation of the simple expansion chamber muffler of Fig. 7.38 for a constant velocity source and zero radiation impedance.

This effect and its remediation may be best illustrated by assuming the source to be a constant velocity source ($Z_s \to \infty$). At such low frequencies, we may make use of the lumped element approximation ($k_0 l, k_0 l_e$ and $k_0 l_t$ being much less than unity) so that $\sin(k_0 l) \to k_0 l$, $\cos(k_0 l) \to 1$), etc., and radiation impedance Z_0 is much less than the tail pipe characteristic impedance $Y_t(= c_0/S_t)$. Then, the exhaust pipe and tail pipe behave like lumped inertance and the expansion chamber acts as lumped compliance (cavity), as shown in Fig. 7.56.

Making use of the linear network theory, it can easily be checked that,

$$\frac{v_0}{v_s} = \frac{Z_c}{Z_c + Z_t} \tag{7.38}$$

where subscripts e, c and t denote exhaust pipe, chamber and tailpipe, respectively.

Substituting Eq. (7.38) in Eq. (7.2) yields,

$$IL = 20 \log \left| \frac{Z_c + Z_t}{Z_c} \right| \tag{7.39}$$

Equation (7.39) shows that $IL \to -\infty$ when $Z_c + Z_t \to 0$. This would happen when [1]:

$$\frac{c_0^2}{j\omega V_c} + j\omega \frac{l_t}{S_t} = 0 \tag{7.40a}$$

where V_c is volume of the chamber (cavity). Equation (7.40a) yields the following frequency for the IL dip in Fig. (7.55).

$$f_{dip} = \frac{\omega}{2\pi} = \frac{c_0}{2\pi} \left(\frac{S_t}{V_c l_t} \right)^{1/2} \tag{7.40b}$$

Substituting $c_0 = 592 \, \text{m/s}$ (corresponding to exhaust gas temperature of $600°C$, $S_t = \pi(0.02)^2 = 0.00126 \, \text{m}^2$, $V_c = \pi * (0.06)^2 * 0.5 = 5.655 \times 10^{-3} \, \text{m}^3$ (5.655 L), Eq. (7.40b) indicates that the IL dip would occur at 81.1 Hz. It may be observed that despite the simplification/assumptions of lumped-element analysis and zero radiation impedance, this value is not far from that in Fig. 7.55.

Substituting Eqs. (7.40a) and (7.40b) in Eq. (7.39) yields the following expression for insertion loss in the neighborhood of the first dip:

$$IL \simeq 20 \log \left| 1 - \omega^2 \frac{l_t V_c}{S_t c_0^2} \right| = 20 \log \left| 1 - \left(\frac{f}{f_{dip}} \right)^2 \right| \simeq 40 \log \left(\frac{f}{f_{dip}} \right) \tag{7.41}$$

Thus, at $f = 2 f_{dip}$, one can expect an IL of about 9.5 dB, as may indeed be verified from Fig. 7.55.

Significantly, the approximate expressions of Eqs. (7.40) and (7.41) throw up the following design guidelines for the engine exhaust muffler, where f_{dip} must be sufficiently less than the firing frequency of the engine:

1. The tail pipe should be long and of narrow cross-section, within the constraints of space and the overriding requirements of low back pressure and aerodynamic noise generation.
2. The muffler proper should have sufficiently large volume. This requirement is reinforced by the need to have large expansion ratio (sharp impedance mismatch). Table 7.1, reproduced here from Refs. [29, 30], gives an idea of how important the muffler volume could be in different octave frequency bands, not just at the low-frequency dip. As a thumb rule, small, medium and large mufflers are roughly characterized by 5, 10 and 15 times the piston displacement capacity of the engine, respectively.

Table 7.1.   Approximate insertion loss (dB) of typical reactive mufflers used with reciprocating engines[a] [29].

Octave band center frequency (Hz)	Low pressure-drop muffler			High pressure-drop muffler		
	Small	Medium	Large	Small	Medium	Large
63	10	15	20	16	20	25
125	15	20	25	21	25	29
250	13	18	23	21	24	29
500	11	16	21	19	22	27
1000	10	15	20	17	20	25
2000	9	14	19	15	19	24
4000	8	13	18	14	18	23
8000	8	13	18	14	17	23

Note: [a]Refer to manufacturers literature for more specific data.

3. We should ensure that exhaust (and intake) system runners are of the same acoustical length so that the firing frequency of the engine is equal to n_{cyl} times that of a single cylinder, where n_{cyl} denotes the number of cylinders — see Eq. (7.33). This would ensure that f_{dip} is much less than (say one-half of) the firing frequency of the engine. If the runner lengths are not equal, as indeed was the case in the older engine designs, then the unmuffled exhaust noise spectrum would show additional low-frequency peaks at the single-cylinder firing frequency (RPM/60 for the 2-stroke cycle engines and RPM/120 for the 4-stroke cycle engines). For such an engine, the muffled exhaust noise would sound considerably louder and harsher (of poor sound quality).

Example 7.5. For a 500 cc, 2-cylinder 4-stroke cycle engine, directly driving an agricultural pump at 3000 RPM, design a single simple expansion chamber muffler (find its volume and the tail pipe diameter and length) so as to ensure an insertion loss of at least 10 dB at the firing frequency of the engine.

Solution. Equations (7.40) and (7.41) indicate that we need to provide large enough expansion chamber volume V_c commensurate with the engine capacity as per Table 7.1, and tail pipe with small enough cross-sectional area S_t and large enough length l_t.

The smallest cross-section of the tail pipe is decided by the consideration of the mean flow Mach number not exceeding 0.2. This would need prior knowledge of the volumetric efficiency and air-fuel ratio of the engine, apart from its capacity, RPM, number of cylinders, cycle-strokes, etc.

For the present problem, let us assume as follows:

$V_c = 10$ times the engine capacity $= 10 \times 500/10^6 = 0.005\,\mathrm{m}^3$

tail pipe diameter $= 2\,\mathrm{cm} \Rightarrow S_t = \frac{\pi}{4} * \left(\frac{2}{100}\right)^2 = 3.1416 \times 10^{-4}\,\mathrm{m}^2$

exhaust gas temperature $= 600°\,\mathrm{C}$, so that

sound speed,

$$c_0 = 592\,\mathrm{m/s}.$$

For a 2-cylinder 4-stroke cycle engine running at $3000\,\mathrm{RPM}$, the firing frequency is given by

$$\omega = 2\pi \frac{3000}{60 * 2} * 2 = 314.16\,\mathrm{rad/s}$$

Substituting these data in Eq. (7.41) yields

$$10 = 20\log \left| 1 - (314.16)^2 \frac{l_t * 0.005}{3.1416 * 10^{-4} * (592)^2} \right|$$

which gives the required tail pipe length: $l_t = 0.928\,\mathrm{m}$, which is reasonable. Thus, an exhaust system with

$$V_c \geq 0.005\,\mathrm{m}^3 = 5\,\mathrm{litres}, \quad d_t \leq 2\,\mathrm{cm} \quad \text{and} \quad l_t \geq 0.928\,\mathrm{m}$$

would ensure an insertion loss of at least $10\,\mathrm{dB}$ at the firing frequency of the given engine.

It may be noted from Eq. (7.41) that insertion loss would improve at lower exhaust temperature and/or with larger expansion chamber or cavity volume.

7.13 Mufflers for High-Pressure Vents and Safety Valves

The foregoing sections have dealt with the intake and exhaust systems of reciprocating engines, fans, blowers, HVAC systems, etc. where the static pressure inside the duct is nearly atmospheric, flow is incompressible, and sound is produced by reciprocating motion of the piston, temporal variation of flow passages and/or periodic cutting of air by rotating blades. By contrast, in the high-pressure vents and safety valves, the static pressure is much above (and often several times) the atmospheric pressure, flow is compressible (choked sonic at the valve throat), and sound is produced by the high-velocity flow of air, gases or superheated steam. High-pressure superheated steam is produced by boilers that take days to stabilize. If, for whatever reasons or eventuality, the steam is not needed for some time, then the boiler cannot be stopped; it keeps producing high-pressure superheated steam, which has to be vented out to the atmosphere. And that produces very high aerodynamic noise.

Another eventuality occurs when due to a malfunction in the line, pressure builds up in the receiver (or plenum) upstream. At a pre-set pressure, a safety valve (or pressure relief valve) opens in order to relieve the extra pressure. This too produces high aerodynamic noise.

To minimize the generation of aerodynamic noise and absorb the noise so produced we use multi-stage diffuser and parallel baffle muffler, respectively. Use of a couple (or more) of multi-hole orifice plates downstream of the valve reduce the high pressure upstream of the valve (in the receiver) in stages. This helps in several ways as follows:

1. Multi-stage expansion (the valve is the first stage in the series) produces much less aerodynamic (flow) noise than a single stage (the valve only).
2. The aerodynamic noise produced by one stage is substantially absorbed by expansion in the following stage(s). Thus, the noise produced at the last stage only needs to be addressed by the parallel baffle muffler.
3. The peak frequency of aerodynamic noise scales with Strouhal number, defined by

$$N_s = \frac{f_d d}{U} \simeq 0.2 \Rightarrow f_p \simeq \frac{0.2U}{d} \tag{7.42}$$

where N_s is Strouhal number, f_p is the peak frequency of the wide-band aerodynamic noise spectrum, U is velocity of the jet emerging from each hole and d is diameter of the hole. As U is near

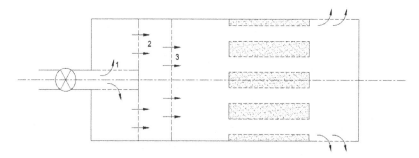

Fig. 7.57   Schematic of a 3-stage diffuser-cum-parallel baffle muffler.

sonic ($\sim$345 m/s) and $d = 5$ mm (or 10 mm), the peak frequency would be of the order of 13,800 (or 6900) Hz. To absorb such high-frequency noise, the required parallel baffle muffler would be of modest dimensions because a parallel baffle muffler is much more efficient at higher frequencies.

Schematic of typical configurations of multi-stage diffuser-cum-parallel baffle muffler is shown in Fig. 7.57.

An alternative to a multi-stage diffuser consisting of multi-hole orifice plates is the use of a multi-expansion trim within the control valve. Some of these high-tech proprietary multiple-step trims are given in Ref. [31] and are reproduced in Chapter 8 (Section 8.11). As the pressure decreases, specific volume increases and therefore the required total throat area would increase with every stage. So, a larger number of holes would be needed as one moves downstream. This calls for a computational fluid dynamics (CFD) analysis of the flow through all the expansion stages including the valve in order to evaluate static pressure, specific volume and the number of holes required at each stage.

For prediction of noise radiated by control valves, the reader is referred to Baumann and Hoffman [32], or to Bies and Hansen [30].

7.14 Design of Muffler Shell and End Plates

In nearly all the muffler configurations discussed above, it is presumed that the muffler shell and end plates are rigid or their

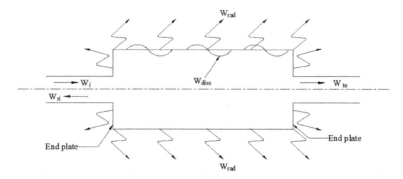

Fig. 7.58   Break-out noise from the muffler shell.

compliance is zero. In actual practice, however, their compliance is non-zero. Excited by the acoustic pressure fluctuations, the shell as well as endplates are set into vibration and radiate secondary sound outside. This is called the break-out noise or the shell noise and is illustrated in Fig. 7.58. This adds up to the noise radiated axially through the tail pipe of the muffler and the net *TL* is then less than the axial *TL* as shown below.

Referring to Fig. 7.58, and neglecting dissipation of power $W_{diss} \to 0$,

Axial transmission loss, $TL_a = 10 \log(W_i/W_{to})$ $\qquad$ (7.43)

Transverse transmission loss, $TL_{tp} = 10 \log(W_i/W_{rad})$ $\qquad$ (7.44)

Net transmission loss, $TL_{net} = 10 \log(W_i/(W_{to} + W_{rad}))$ $\qquad$ (7.45)

where

W_i is the incident acoustic power in the exhaust pipe,

W_{ri} is the reflected acoustic power in the exhaust pipe,

W_{to} is the acoustic power transmitted into an anechoic termination along the tail pipe,

W_{rad} is the acoustic power radiated as break-out noise from the shell and end plates, and

W_{diss} is the power that is dissipated into heat through viscosity or structural damping.

Substituting Eqs. (7.43) and (7.44) into Eq. (7.45) yields

$$TL_{net} = -10\log(10^{-0.1\,TL_a} + 10^{-0.1\,TL_{tp}}) \qquad (7.46)$$

Equation (7.46) shows that TL_{net} at any frequency would be less than the lower of TL_a and TL_{tp} by 0 to 3 dB.

Unlike axial transmission loss, TL_a, which is zero as frequency f tends to zero, the transverse transmission loss (TTL), TL_{tp} tends to infinity as f tends to zero, as is typical of the stiffness controlled TL. TL_{tp} is maximum for an ideal circular cylinder and falls drastically with non-circularity [33], as is shown in Fig. 7.59. Deviation from circularity decreases the TTL and hence increases the break-out noise substantially [34]. That is why TTL of rectangular (or square) ducts is the lowest [35], which accounts for "cross-talk"' in the HVAC systems. This indicates the importance of proper design of shell and end plates [36] as well as the inner acoustic elements of the muffler.

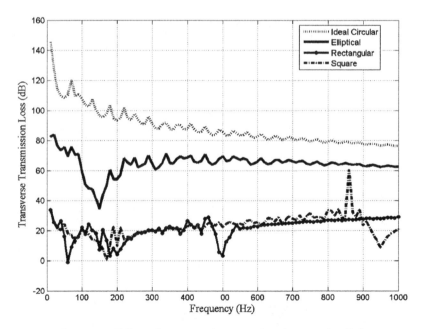

Fig. 7.59 Effect of non-circularity on break-out noise [34].

The break-out noise from the end plates may be reduced by stiffening them and giving them a dish shape; flat plates act as efficient radiators. Offset inlet and outlet automatically serve to stiffen the end plates and should be preferred in order to reduce the break-out noise.

7.15 Helmholtz Resonators

The resonator developed by Helmholtz is as old as the science of Acoustics. It consists of a cavity (lumped volume) attached to a duct through a small narrow tube (neck) as shown in Fig. 7.60(a). Its analogous circuit presentation is shown in Fig. 7.60(b), where it is represented as a branch or shunt impedance consisting of a lumped inertance due to the neck and compliance due to the cavity. Additionally, there is radiation impedance at both ends of the neck.

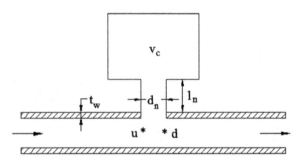

(a) Schematic of Helmhaltz resonator.

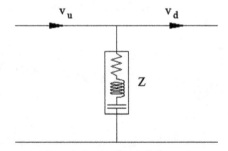

(b) Analogous circuit of Helmholtz resonator

Fig. 7.60   Lumped-element modeling of HR in a duct.

Thus, the impedance of the resonator is given by [1]:

$$Z = \frac{\omega^2}{\pi c_0} + j\omega \frac{l_{ne}}{S_n} + \frac{c_0^2}{j\omega V_c} \qquad (7.47a)$$

where S_n is the neck cross-sectional area, $\pi d_n^{2/4}$, l_{ne} is the effective length of the neck given by

$$l_{ne} = l_n + t_w + 0.85 d_n \qquad (7.47b)$$

Here, the end correction $0.85 d_n$ represents the sum of the inertive parts of the radiation impedance at the two ends of the neck.

At the resonance frequency of the resonator, the reactive (imaginary) part of Z would tend to zero. Thus, the resonance frequency of Helmholtz resonator (HR) is given by

$$j\omega \frac{l_{ne}}{S_n} + \frac{c_0^2}{j\omega V_c} = 0$$

which, on rearrangement, yields

$$f_n = \frac{\omega}{2\pi} = \frac{c_0}{2\pi} \left(\frac{S_n}{l_{ne} V_c} \right)^{1/2} \qquad (7.48)$$

At this frequency, most of the incoming acoustic power would be reflected back to the source because of the acoustical short-circuit. At this frequency, TL of the HR would tend to infinity. Thus, the TL curve would show a sharp peak — the resonance peak of the HR.

The half-power bandwidth of the resonance peak in the TL curve would tend to zero if the resistive part of the resonator impedance in Eq. (7.47) were zero. However, the term $\omega^2/\pi c_0$ gives a finite width and amplitude to the resonance peak. The half-power bandwidth can be increased at the cost of its amplitude by lining the cavity acoustically. The acoustical, as well as anti-drumming lining on the inside of the cavity, helps to reduce the structural vibration of the cavity and the consequent break-out noise.

HRs are characterized by high TL at the resonance frequency. They are ineffective at all other frequencies except in the immediate neighborhood of the resonance frequency. Therefore, they find application on constant-speed machines like generator sets (gensets).

They are also used sometimes on the intake systems of automotive engines in order to suppress a boom in the passenger cabin at the idling speed of the engine.

Another application of HR or its uniform-area transverse tube variant is on the intake and/or discharge duct of the constant-speed fans or blowers. As the unmuffled intake and exhaust-noise spectra are characterized by sharp peaks at the blade passing frequency (BPF) and its first few multiples, one can design and employ transverse tubes resonating at the BPF, 2 times BPF and 3 times BPF as shown in Fig. 7.61.

The branch impedance of transverse uniform tube of length l and area of cross-section S is given by

$$Z = -jY_0 \cot k_0 l, Y_0 = c_0/S \qquad (7.49)$$

This would tend to zero when,

$$k_0 l = (2m - 1)\pi/2, \quad k_0 = \omega/c_0 = 2\pi f_m/c_0 \qquad (7.50a)$$

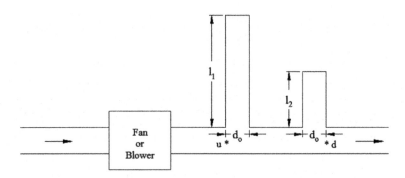

(a) Schematic of the configuration with two resonators.

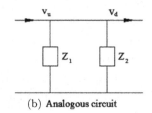

(b) Analogous circuit

Fig. 7.61 Transverse tube resonators for the exhaust noise control of a constant speed fan or blower.

or when the frequency equals one of the resonances frequencies:

$$f_m = (2m - 1)\frac{c_0}{4l}, \quad m = 1, 2, 3, \ldots \quad (7.50b)$$

Thus, if $l = c_0/(4 \times BPF)$, then it would suppress the BPF peak and its odd multiples. Similarly, a transverse tube of half the length would suppress the SPL peaks occurring at even multiples of the BPF. Hence, referring to Fig. 7.61, and incorporating the end corrections, the design values of l_1 and l_2 work out to be as follows:

$$l_1 = \frac{c_0}{4 * BPF} - 0.425 \, d_0 \quad (7.51)$$

$$l_2 = \frac{c_0}{8 * BPF} - 0.425 \, d_0 \quad (7.52)$$

Example 7.6. A tube-axial fan has eight blades and is running at a constant speed of 1500 RPM. Assuming the average ambient temperature of 25°C, design the quarter-wave resonators in order to suppress the fan noise at its BPF and its first two harmonics.

Solution. Blade passing frequency of the fan,

$$BPF = \frac{8 * 1500}{60} = 200 \, \text{Hz}$$

At 25° C, source speed $c_0 = 346 \, \text{m/s}$.

Let the transverse quarter-wave resonator tube diameter, $d_0 = 0.1 \, \text{m}$.

Then, making use of Eqs. (7.51) and (7.52), the required lengths of the resonators (see Fig. 7.61) are given by

$$l_1 = \frac{346}{4 * 200} - 0.425 * 0.1 = 0.39 \, \text{m} = 390 \, \text{mm}$$

$$l_2 = \frac{346}{8 * 200} - 0.425 * 0.1 = 0.174 \, \text{m} = 174 \, \text{mm}$$

The resonator tube of 390 mm length would absorb BPF and its 3rd, 5th, ... multiples, and the tube of 174 mm length would absorb 2nd, 6th, ... multiples of the BPF. Therefore, two resonator tubes of length 390 mm and 174 mm would suffice to suppress the BPF and its first two harmonics.

There is another interesting application of HRs. When a large number of identical HR's are applied over a wall, they can make the wall anechoic to plane wave incident on the wall when the plane wave frequency equals the resonance frequency of the HR's. At resonance, impedance Z of Eq. (7.47) reduces to pure resistance, $R = \omega^2/\pi c_0$. This resistance would absorb acoustic power equal to $p_{rms}^2/(\rho_0 R)$. Now, power in a plane wave normal to area A is $p_{rms}^2 A/\rho_0 c_0$. Equating the two gives an expression for area A over which the wall will act as anechoic. Thus,

$$\frac{p_{rms}^2 A}{\rho_0 c_0} = \frac{p_{rms}^2}{\rho_0 R} \tag{7.53}$$

or

$$A = \frac{c_0}{R} = \frac{c_0}{\omega^2/\pi c_0} = \frac{\pi c_0^2}{\omega^2} \approx \frac{0.953}{(f/100)^2}\,(\mathrm{m}^2) \tag{7.54}$$

Thus, at room temperature of 25°C ($c_0 = 346\,\mathrm{m/s}$), an HR resonating at 100 Hz would make 0.953 m² area of the wall anechoic at that frequency.

7.16 Active Noise Control in a Duct

An ANC system makes use of a secondary sound source and an adaptive digital controller to generate a pressure signal that is out of phase with that of the primary sound source. Schematic of an ANC system in a duct is shown in Fig. 7.62. Signal conditioner shown there may consist of a preamplifier, LPF and analog-to-digital (A/D) converter. Secondary source of sound is generally a loudspeaker specially designed for the purpose. The error signal is used to adapt the filter of the microprocessor in order to generate the required transfer function that would help to minimize acoustic pressure at the error microphone.

The acoustic signal produced by the secondary source travels upstream as well as downstream and is picked up (sensed) by the

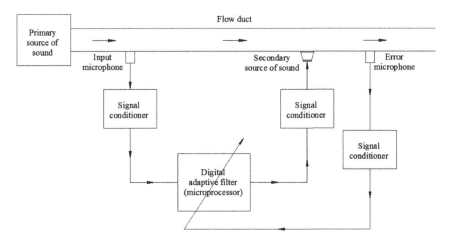

Fig. 7.62   Schematic of an ANC system in a duct.

input microphone as well as the error microphone. It has been proved
by means of a 1D standing wave analysis [37] as well as the progressive wave transfer function analysis [38] that in an ANC system in its
steady state, the primary source faces nearly zero acoustic resistance,
and so does the secondary source. In other words, the two sources
unload each other. The controller helps to create an acoustical short
circuit at the secondary source junction. When the acoustic load
(resistance) at each source is zero, the particle (or volume) velocity
is out of phase with pressure, and therefore, does not produce any
audible sound. In other words, the primary source and secondary
source muffle each other. Thus, an ANC system produces a sharp
impedance mismatch within a limited frequency band [37–41].

Adaptive controllers are basically of two types: feedforward controller and feedback controller. Feedforward controller works efficiently for periodic noise (from fans, blowers, HVAC systems that
run at constant speed) or for random (turbulence-generated) noise
propagating in ducts. The feedback controller works best for active
earmuffs (headsets), active vehicle suspension systems and active
control of structural vibration. Analysis and design of controllers call
for extensive knowledge of digital signal processing (DSP), which is
out of the scope of the textbook.

A typical engine exhaust system or industrial fan produces a signal that varies with time because of small (but unavoidable) variations in speed, and therefore adaptive controller is a necessity for an ANC system in a duct.

Large (powerful) loudspeaker(s), signal conditioners and adaptive DSP controller make an ANC a costly proposition. However, such a system has the advantage of minimal stagnation pressure drop which reduces the running cost of the system dramatically. Thus, for large industrial fans, the ANC system reduces the power bill substantially. It not only offsets the disadvantage of high initial hardware cost, but also offers a very cost-effective alternative to passive noise control system.

ANC system in a duct works best for plane wave propagation. The presence of higher-order modes at high frequencies would call for more than one input microphone, secondary source loudspeakers and error phones, apart from a more complex DSP controller. This would make the ANC system uneconomical at higher frequencies. For the same reason, ANC system is uneconomical for use in 3D enclosures or rooms for global noise control. However, ANC system can be used on seats of luxury vehicles, aircraft passenger cabins, etc. for localized noise control, roughly within a radius of quarter wavelength.

ANC system is most effective at lower frequencies (lower than 1000 Hz, rather 500 Hz) where passive noise control system, particularly the absorptive type, is not very effective. This makes the ANC system complementary to the passive noise control system for wideband noise control requirements.

The ANC system, when tuned adaptively, locks on to the peaks in the unmuffled noise spectrum, and chops them off, as it were. This is what makes the ANC system particularly suitable for large fans and blowers, the noise spectrum of which is characterized by sharp peaks at the BPF and its first few harmonics.

7.17 Pressure Drop Considerations

Mean flow through the muffler passages encounters a drop in its stagnation pressure. The engine or fan source must push the gases or air against this back pressure. This results in pumping losses in the engine power output. When back pressure is high, it results in a drop in volumetric efficiency, which in turn would result in a further loss in engine shaft power and an increase in the specific fuel consumption (SFC) of the engine. In the case of a fan, the power load on the motor is proportional to the product of the flow rate Q and total pressure drop, Δp. The latter is the sum of the pressure drops in the intake system and discharge/exhaust system. Thus, electrical power consumption of a fan motor is directly proportional to the total back pressure experienced by the fan. Therefore, prediction and minimization of the stagnation pressure drop across mufflers (or silencers) is an integral part of the discipline of analysis and design of mufflers.

Two primary mechanisms of the stagnation pressure drop are wall friction in the boundary layer, and eddy and vortex formation in the free shear layer at area discontinuities and perforated sections.

Pressure drop can be normalized in terms of dynamic head, H. Thus,

$$\Delta p = K.H, \quad H \equiv \frac{1}{2}\rho_0 U^2 \qquad (7.55)$$

Here, K is called the dynamic pressure loss factor or coefficient. Empirical expressions of K for different muffler elements with turbulent mean flow are given as follows.

Pressure loss factor for turbulent flow through a pipe or duct of length l_p and (hydraulic) diameter d_p (Fig. 7.3):

$$K \equiv \frac{\Delta p}{H} = F \cdot \frac{l_p}{d_p} \qquad (7.56)$$

where F is the Froude's friction factor given by

$$F = 0.0072 + 0.612/R_e^{0.35}, \quad R_e < 4 \times 10^5 \qquad (7.57)$$

$R_e = U d_p \rho_0 / \mu$ is the Reynold's number, and μ and p_0 are the coefficient of dynamic viscosity and mass density, respectively.

Typically, $F \approx 0.016$ for the mild steel pipes used in exhaust mufflers. However, roughness of fiberglass-lined pipes is much more than that of the normal unlined rigid pipes, and therefore, $F \approx 0.032$

for the lined pipes as well as the perforated unlined pipes, as in a CTR (Fig. 7.10).

Pressure loss factor for a sudden contraction (Fig. 7.4):

$$K \simeq (1 - n)/2, \quad n = (r_d/r_u)^2 \tag{7.58}$$

Pressure loss factor for a sudden expansion (Fig. 7.5):

$$K \simeq (1 - n)^2, \quad n = \left(\frac{r_u}{r_d}\right)^2 \tag{7.59}$$

Note that for both the sudden area discontinuities, n is defined as ratio of cross-section of the narrower pipe to that of the wider pipe. Thus, n is less than unity in both Eqs. (7.58) and (7.59).

Pressure loss factor for extended outlet (Fig. 7.6), extended inlet (Fig. 7.7), flow reversal with contraction or expansion:

$$K \simeq 1.0 \tag{7.60}$$

Pressure loss factors for bell mouth, gradual contraction, gradual expansion, etc. are much lower than those for the corresponding sudden or step area changes. However, gradual area changes lead to poor *TL* and therefore are not always desirable.

Pressure loss coefficient of the plug muffler (Fig. 7.44) [12]:

$$K = 5.6e^{-0.23x} + 67.3e^{-3.05x}, \quad 0.25 < x < 1.4 \tag{7.61}$$

where x is the open-area ratio of the crossflow expansion (Fig. 7.11) and crossflow contraction (Fig. 7.12), assumed to be equal. The open-area ratio is defined by

$$\text{open-area ratio, } x = \frac{2\pi r_1 l_p \sigma}{\pi r_1^2} = \frac{2 l_p \sigma}{r_1} \tag{7.62}$$

where σ is porosity of each of the two perforated sections.

Pressure loss coefficient of the muffler element with three interacting ducts (Fig. 7.13) [12]:

$$K = 4.2e^{-0.06x} + 16.7e^{-2.03x} \tag{7.63}$$

By contrast, pressure loss coefficient of a CTR (Fig. 7.10) is as small as [12]:

$$K = 0.06x \qquad (7.64)$$

where x is the open-area ratio defined by Eq. (7.62).

Example 7.7. For length $l = 0.5\,\text{m}$, $r = 2\,\text{cm}$, $R = 6\,\text{cm}$ estimate the stagnation pressure drop across the simple expansion chamber muffler of Fig. 7.38.

Solution. Area ratio,

$$n = (r/R)^2 = (2/6)^2 = 0.111$$

As per Eq. (7.59), the pressure loss coefficient for sudden expansion,

$$K_e = (1 - n)^2 = (1 - 0.111)^2 = 0.790.$$

As per Eq. (7.56), pressure loss coefficient for wall friction to turbulent flow through the chamber length,

$$K_f = \frac{Fl}{D} \simeq 0.016 \left(\frac{0.5}{2 * 0.06} \right) = 0.067$$

As per Eq. (7.58), pressure loss coefficient for sudden contraction,

$$K_c = (1 - n)/2 = (1 - 0.111)/2 = 0.444$$

Let the dynamic head in the exhaust (and tail) pipe, $H = \frac{1}{2}\rho_0 U^2$
Then, flow velocity in the chamber $= U \times n = 0.111U$.
And therefore, dynamic head in the chamber $= (0.111)^2 H = 0.0123\,H$.

Thus, the total stagnation pressure drop across the muffler proper (excluding the exhaust pipe and tail pipe) is given by

$$\Delta_p = (0.790 + 0.0123 \times 0.067 + 0.444)\,H$$

$$= (0.790 + 0.0008 + 0.444)\,H = 1.235\,H$$

Hence, the total stagnation pressure drop is equal to 1.235 times the dynamic head in the exhaust/tail pipe.

It may be noted that the contribution of the chamber to the total pressure drop is very small, and therefore can be neglected for most design calculations.

For multiply connected perforated element mufflers, one needs to make use of the lumped flow resistance network circuit [15] in order to predict the pressure loss coefficient of the muffler. Fortunately, suchlike mufflers are characterized by quite small pressure drop, and therefore represent a viable design option.

The foregoing expressions for the pressure loss coefficient are approximate. Therefore, ultimately, the overall back pressure of the muffler has to be measured on the test bed.

In the case of automotive engines, the after-treatment devices like catalytic converter, diesel particulate filter, etc., add substantially to the overall back pressure experienced by the engine. That leaves only a modest pressure drop for which the muffler needs to be designed. In fact, the maximum permissible pressure drop across the muffler is the most important consideration for synthesizing an appropriate muffler configuration.

References

[1] Munjal, M. L., *Acoustics of Ducts and Mufflers*, Second Edition, Wiley, Chichester, UK (2014).

[2] Munjal, M. L., *Muffler acoustics*, Chapter K, in *Formulas of Acoustics* (Ed. Mechel, F.P.), Springer Verlag, Berlin (2006).

[3] Munjal, M. L., Sreenath, A. V. and Narasimhan, M. V., A rational approach to the synthesis of one-dimensional acoustic filters, *Journal of Sound and Vibration*, 29(3), pp. 263–280 (1973).

[4] Munjal, M. L., Galaitsis, A. G. and Ver, I. L., Passive silencers, Chapter 9, in (Eds. Ver, I. L. and Beranek, L. L.), *Noise and Vibration Control Engineering*, John Wiley, New York (2006).

[5] Hans Boden, and Ragnar Glav, Exhaust and intake noise and acoustical design of mufflers and silencers, Chapter 85 in *Handbook of Noise and Vibration Control* (Ed. Crocker, M. J.), John Wiley, New York (2007).

[6] Kinsler, L. E., Frey, A. R., Coppens, A. B. and Sanders, J. V., *Fundamentals of Acoustics*, Fourth Edition, John Wiley, New York (2000).

[7] Chaitanya, P. and Munjal, M. L., Effect of wall thickness on the end-corrections of the extended inlet and outlet of a double-tuned expansion chamber, *Applied Acoustics*, 72(1), pp. 65–70 (2011).

[8] Karal, F.C., The analogous impedance for discontinuities and constrictions of circular cross-section, *Journal of the Acoustical Society of America*, 25(2), pp. 327–334 (1953).

[9] Sahasrabudhe, A. D., Munjal, M. L. and Ramu, P. A., Analysis of inertance due to the higher order mode effects in a sudden area discontinuity, *Journal of Sound and Vibration*, 185(3), pp. 515–529 (1995).

[10] Sullivan, J. W. and Crocker, M. J., Analysis of concentric tube resonators having unpartitioned cavities, *Journal of the Acoustical Society of America*, 64, pp. 207–215 (1978).

[11] Chaitanya, P. and Munjal, M. L., Tuning of the extended concentric tube resonators, *International Journal of Acoustics and Vibration*, 16(3), pp. 111–118 (2011).

[12] Munjal, M. L., Krishnan, S. and Reddy, M. M., Flow-acoustic performance of the perforated elements with application to design, *Noise Control Engineering Journal*, 40(1), pp. 159–167 (1993).

[13] Mimani, A. and Munjal, M. L., Transverse plane-wave analysis of short elliptical end-chamber and expansion chamber mufflers, *International Journal of Acoustics and Vibration*, 15(1), pp. 24–38 (2010).

[14] Mimani, A. and Munjal, M. L., Transverse plane-wave analysis of short elliptical chamber mufflers — An analytical approach, *Journal of Sound and Vibration*, 330, pp. 1472–1489 (2011).

[15] Elnady, T., Abom, M. and Allam, S., Modeling perforate in mufflers using two-ports, *ASME Journal of Vibration and Acoustics*, 132(6), pp. 1–11 (2010).

[16] Panigrahi, S. N. and Munjal, M. L., Plane wave propagation in generalized multiply connected acoustic filters, *Journal of the Acoustical Society of America*, 118(5), pp. 2860–2868 (2005).

[17] ASHRAE (American Society of Heating, Refrigeration and Air-Conditioning Engineers), *Handbook: HVAC Applications*, Chapter 46 (1999).

[18] Panigrahi, S. N. and Munjal, M. L., Combination mufflers — Theory and parametric study, *Noise Control Engineering Journal*, 53(6), pp. 247–255 (2005).

[19] Prasad, M. G. and Crocker, M. J., On the measurement of the internal source impedance of a multi-cylinder engine exhaust system, *Journal of Sound and Vibration*, 90, pp. 491–508 (1983).

[20] Callow, G. D. and Peat, K. S., Insertion loss of engine inflow and exhaust silencers, *I Mech. E* C19/88, pp. 39–46 (1988).

[21] Hota, R. N. and Munjal, M. L., Approximate empirical expressions for the aeroacoustic source strength level of the exhaust system of compression ignition engines, *International Journal of Aeroacoustics*, 7(3&4), pp. 349–371 (2008).

[22] Fairbrother, R., Boden, H. and Glav, R., Linear acoustic exhaust system simulation using source data from linear simulation, *SAE Technical Paper Series*, 2005-01-2358 (2005).

[23] Boden, H., Tonse, M. and Fairbrother, R., On extraction of IC-engine acoustic source data from non-linear simulations, in *Proceedings of the Eleventh International Congress on Sound and Vibration* (ICSV11), St. Petersburg, Russia (2004).

[24] Knutsson, M. and Boden, H., IC-Engine intake source data from non-linear simulations, SAE Technical Paper Series, 2007-01-2209 (2007).

[25] Hota, R. N. and Munjal, M. L., Intake source characterization of a compression ignition engine: Empirical expressions, *Noise Control Engineering Journal*, 56(2), pp. 92–106 (2008).

[26] Munjal, M. L. and Hota, R. N., Acoustic source characteristics of the exhaust and intake systems of a spark ignition engine, Paper No. 78, Inter-Noise 2010, Lisbon, Portugal (2010).

[27] BOOST Version 7.0.2, AVL LIST GmbH, Graz, Austria (2007).

[28] Munjal, M. L., Panigrahi, S. N. and Hota, R. N., FRITAmuff: A comprehensive platform for prediction of unmuffled and muffled exhaust noise of I.C. engines, in *14th International Congress on Sound and Vibration* (ICSV14), Cairns, Australia (2007).

[29] Joint Departments of the Army, Air Force and Navy, USA, Power Plant Acoustics, Technical manual TM 5-805-9 AFT 88-20 NAVFAC Dhf 3.14 (1983).

[30] Bies, D. A. and Hansen, C. H., *Engineering Noise Control: Theory and Practice*, Fourth Edition, Spon Press, London (2009).

[31] Baumann, H. D. and Coney, W. B., Noise of gas flows, Chapter 15, in *Noise and Vibration Control* (Eds. Ver, I. L. and Beranek, L. L.), Second Edition, John Wiley, New York (2006).

[32] International Standard Industrial-process control valves — Part 8-3: Noise considerations — Control valve aerodynamic noise prediction method, IEC 60534-3-3 Edition 3.0 (2010-11).

[33] Cummings, A., Chang, I.-J. and Astley, R. J., Sound transmission at low frequencies through the walls of distorted circular ducts, *Journal of Sound and Vibration*, 97, pp. 261–286 (1984).

[34] Munjal, M. L., Gowtham, G. S. H., Venkatesham, B. and Harikrishna Reddy, H., Prediction of breakout noise from an elliptical duct of finite length, *Noise Control Engineering Journal*, 58(3), pp. 319–327 (2010).

[35] Venkatesham, B., Pathak, A. G. and Munjal, M. L., A one-dimensional model for prediction of breakout noise from a finite rectangular duct with different acoustic boundary conditions, *International Journal of Acoustics and Vibration*, 12(3), pp. 91–98 (2007).

[36] Narayana, T. S. S. and Munjal, M. L., Computational prediction and measurement of break-out noise of mufflers, in *SAE Conference, SIAT 2007*, SAE Paper 2007-26-040, ARAI, Pune, India (2007), pp. 501–508.

[37] Munjal, M. L. and Eriksson, L. J., An analytical, one-dimensional standing wave model of a linear active noise control system in a duct, *Journal of the Acoustical Society of America*, 84(3), pp. 1086–1093 (September 1988).

[38] Munjal, M. L. and Eriksson, L. J., Analysis of a linear one-dimensional noise control system by means of block diagrams and transfer functions, *Journal of Sound and Vibration*, 129(3), pp. 443–455 (March 1989).

[39] Hansen, C. H. and Snyder, S. D., *Active Control of Sound and Vibration*, E&FN Spon, London (1997).

[40] Hansen, C. H., *Understanding Active Noise Cancellation*, E&FN Spon, London (2001).

[41] Nelson, P. A. and Elliott, S. J., *Active Control of Sound*, Academic Press, London (1992).

Problems

7.1. Design a simple expansion chamber muffler for a *TL* peak of 15 dB at a frequency of 125 Hz, for plane waves in stationary medium at 25°C.

[Ans.: Chamber length, $l = 0.692$ m; area ratio, $m = 11.16$]

7.2. Design a simple expansion chamber muffler for *TL* of at least 10 dB from 50 to 100 Hz for stationary medium at 25°C.

[Ans.: Chamber length, $l = 1.153$ m; area ratio, $m = 7.07$]

7.3. Design a circular, double-tuned ETEC (find out the extended inlet and outlet lengths, l_a and l_b) of length 0.5 m, area ratio 9.0, inlet/outlet tube diameter and thickness equal to 40 mm and 1 mm, respectively, for stationary medium at 25°C.

[Ans.: $l_{g,a} = 241.5$ mm and $l_{g,b} = 116.5$ mm]

7.4. Design a tuned extended-tube concentric tube resonator within an axi-symmetric chamber with $l = 0.5$ m, $d = 40$ mm, $D = 120$ mm, perforate porosity, $\sigma = 0.1$.

[Ans.: $l_{g,a} = 232.4$ mm and $l_{g,b} = 107.4$ mm]

7.5. A 2 m long, 800 mm × 800 mm square duct with rigid walls is proposed to be lined on all four sides (see image (a) in the following figure) with 100 mm thick mineral wool blanket with sound absorption coefficient of 0.6 in the 250 Hz octave band.

Evaluate its *TL*. If the lining were re-configured so as to form a 4-pass parallel baffle muffler, as shown in the following figure (b), what would its *TL*?

Assume that the sound absorption coefficient of the 50 mm thick blanket is 0.3 in the 250 Hz band.

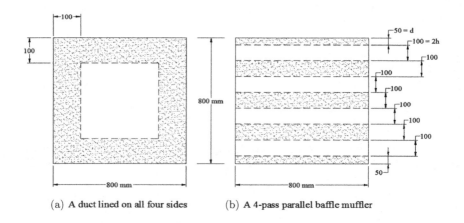

(a) A duct lined on all four sides   (b) A 4-pass parallel baffle muffler

[Ans.: 12 dB; 18 dB]

7.6. Design a muffler (estimate the chamber volume and length of the 6 mm diameter tail pipe) for a single-cylinder, 50 cc 4-stroke-cycle engine running at 6000 RPM so as to ensure insertion loss of 8 dB at the firing frequency. Assume the speed of sound in the exhaust gases to be 500 m/s.

[Ans.: Chamber volume, $V_c = 500$ cc; length of the tail pipe, $l_t = 0.503$ m; Alternatively, $V_c = 1$ L and $l_t = 252$ mm]

7.7. Evaluate length of the 2.4 mm-diameter neck of a HR opening into a cavity of 1 L volume for the intake system of a 4-cylinder 4-stroke diesel engine in order to suppress the firing frequency tone of the engine idling at 900 RPM. State the assumptions made in the process.

[Ans.: Neck length, $l_n = 11.95$ mm]

7.8. Evaluate and compare the back pressure of the following four muffler configurations:

(a) simple expansion chamber (Fig. 7.38),
(b) concentric tube resonator (Fig. 7.10),
(c) crossflow, 3-duct closed-end element (Fig. 7.13),
(d) plug muffler (Fig. 7.44).

For all the four configurations, diameter of the exhaust pipe as well as tail pipe is 40 mm; diameter of the circular shell is 160 mm; porosity and length of the perforates (in configurations b, c and d) are 0.05 and 0.3 m, respectively; the mean flow velocity in the exhaust pipe (as well as tail pipe) is 50 m/s and temperature of the medium is 500°C.

[Ans.: (a) 770 Pa, (b) 51.4 Pa, (c) 2646 Pa,
(d) 2679 Pa]

Chapter 8

Noise Control Strategies

8.1 Introduction

Noise may be controlled at the source, in the path or at the receiver end. The cost of noise control increases as one moves away from the source. In fact, designing for quietness is the most cost-effective way; prevention is better than cure! In this chapter, we list and illustrate all possible techniques and strategies for noise control of an existing machine as well as choice or design of quieter machines, processes and industrial layouts.

8.2 Control of Noise at the Source

Noise of a machine may be reduced at the source by adopting some of the following measures or practices.

8.2.1 *Select a quieter machine*

Select a quieter machine from the market even if it is relatively costlier. Often, this additional cost is less than the cost and hassle of the retrofit noise control measures. A machine would generally be quieter if its moving parts were fabricated to closer tolerances. Manufacturing mating parts to closer tolerances reduces micro-impacting between the parts. This in turn

- reduces the mechanical (impact) noise,
- reduces vibrations,
- reduces wear and tear,
- increases fatigue life,

- increases the interval between maintenance outings, and
- increases the accuracy of a machine tool.

Overall, manufacturing to closer tolerances is a good engineering practice and makes economic sense in the long run. In other words, the lifetime cost of a quieter, though costlier, machine is relatively lower.

Human desire to obtain large power from small power packs, particularly for aeroengines, motorbikes and automobiles, has resulted in a race for high-speed engines and turbines. However, noise control at the source can be achieved by selecting large, slow machines rather than small, faster ones, particularly for stationary installations of captive diesel generators, compressors, etc.

8.2.2 *Select lossy materials*

Mechanical impacting sets different parts of a machine into free vibration. The sheet metal components or covers, vibrating in their natural modes, radiate noise into the atmosphere. Free vibration of sheet metal components can be reduced by fabricating these components out of lossy materials like plastics; that is, materials with relatively high loss factor. We could consider replacing metal gears with plastic gears. Of course, such materials are very low in strength, and therefore, cannot be used in high-stress locations. Alternatively, we could make use of free layer damping (FLD), or preferably, constrained layer damping (CLD) (discussed in Chapter 4) for the sheet metal components or covers exposed to the atmosphere. For example, circular saw blades should be replaced with damped blades.

8.2.3 *Use quieter processes or tools*

One very popular way of reducing noise in energy conversion is the use of electrical motors rather than reciprocating engines or gas turbines.

Use of high velocity jets for cleaning must be avoided as far as possible in industry. We should seriously consider replacing pneumatic tools with electric tools, and pneumatic ejectors with mechanical ejectors.

Forging down to the net shape in a single step die is another source of high noise in workshops. We must consider replacing single operation dies with stepped dies, and rolling or forging with pressing.

Increasing the contact time during impacting reduces the rate of change of force. Therefore, rotating shears produce less noise than mechanical presses, which in turn are generally quieter than hammers. Welding or squeeze riveting creates less noise than impact riveting.

Increasing the contact area is a potent method for reducing noise at the source in material handling. Therefore, use of belt conveyers instead of roller conveyers recommends itself. For the same reason, helical gears have been observed to produce much less noise than spur gears, and spiral cutters are quieter than straight-edge cutters by upto 10 dB. Probably, the best example of noise reduction by increasing the contact area is use of belt drives in place of gear drives, and pneumatic tyres instead of solid steel wheels. As a matter of fact, development of pneumatic tyres made of rubber and nylon strings was probably the most important step towards reducing automobile noise and preserving the pavement.

8.2.4 *Reduce radiation efficiency*

For a given vibration amplitude, noise radiation from a structure depends upon radiation efficiency, which in turn depends on the shape and texture of the vibrating surfaces. Flat continuous plates are efficient radiators of sound, and therefore must be replaced with curved or ribbed panels, perforated plates, or woven strips of metal. Two sides of a plate radiate out-of-phase waves, and perforations expose both sides of the plate simultaneously, resulting in substantial cancellations at the holes or periodic gaps. This effect roughly represents conversion of monopole mechanism of noise generation into dipole mechanism, which has intrinsically much lower radiation efficiency.

This mechanism of noise reduction can be, and often is, utilized in design of hoppers, stillages or tote boxes for material handling operations. Often, the noise in a factory is increased dramatically when the metal ore, castings or waste material is tipped into a tote box made of sheet metal. If this tote box or stillage were made of interwoven metal strips, noise reduction of the order of 20 dB could be obtained. This represents a tremendous noise reduction at the source at minimal cost. This principle can also be applied to cover transmission couplings, baggage trolleys on railway platforms, side covers of mechanical hammers and presses, etc.

8.2.5 *Maintain for quietness*

Often, it is observed that a machine that has been in use for a while is no longer as quiet as it was when it was first installed. In order to avoid this additional noise at the source, it is necessary to:

(a) have the rotating parts of the machine balanced on site, not at the supplier's premises,
(b) monitor the condition of the bearings continuously and have them lubricated regularly,
(c) replace or adjust the worn or loose parts as soon as detected, and
(d) follow the periodic maintenance schedules specified by the supplier.

If a machine is thus maintained, its noise level would not unduly increase with use or age. As indicated elsewhere in this book, increased noise is often a symptom of poor maintenance of the machine. If not attended to, the primary function of the machine would be compromised and the chances of its failure or malfunction would increase alarmingly.

8.2.6 *Design of flow machinery for quietness*

Most often, noise is radiated by cooling equipments. As a rough estimate, half of all noise in the world emanates from fans, blowers, and heating ventilating and air-conditioning (HVAC) systems. Noise control of flow machinery at the source may be affected as follows:

(a) Provision of adequate cooling fins would reduce or eliminate the need of fan for forced convection cooling. Then, either the fan would not be needed at all, or a small, relatively quieter, fan would do.

(b) As will become clear later in this chapter, centrifugal fans are generally quieter than propeller fans of the same capacity working against the same back pressure. Again, backward-curved blades produce less noise than forward-curved blades. Computational fluid dynamics (CFD) can be used to design quieter blade profiles like airfoil blades.

(c) If the flow entering a fan does not have uniform velocity profile, or is highly turbulent, then the fan would produce more noise than usual. Therefore, we should locate the fans in smooth, undisturbed air flow, not too near a bend, for example.

(d) Large low-speed fans are generally quieter than smaller faster ones.

(e) For shop floor cooling, it is better to provide each worker with small individual fans than installing large high-speed fans at one end of the workshop.

(f) Noise in high-velocity flow ducts increases substantially at the sharp bends due to separation of boundary layer. Therefore, adequate CFD modeling may be done, and suitable guide vanes may be provided to avoid flow separation.

(g) Flow noise or aerodynamic noise from subsonic jets increases with sixth to eighth power of flow velocity. Therefore, we should design the flow ducts so as to maximize the cross-section of flow streams. Incidentally, this consideration led to the development and use of high bypass turbofan engines for transport aircraft. It reduced the jet noise by 10–20 dB while retaining the same thrust. Moreover, propulsion efficiency of turbofan engines is higher than the corresponding turbojet engines.

(h) Drastic reduction of pressure (as in high-pressure venting of air or superheated steam) in a single step (single-stage expansion trim valve) produces very high sound pressure levels. As explained before in Chapter 7, this may be reduced by means of a multistage expansion trim valve or a series of adequately designed orifice plates.

(i) As will be learnt later in this chapter, noise of a fan or blower increases with flow rate as well as the back pressure against which the flow needs to be pushed. Therefore, the flow passages need to be designed so as to minimize the stagnation pressure drop in the system. Incidentally, it will also reduce the load on the fan and hence the power consumption.

(j) Often, the nozzles used in high-pressure pneumatic cleaning devices produce high jet noise. This can be reduced by means of specially designed multi jet nozzles or by-pass nozzles.

8.3 Control of Noise in the Path

For any existing noisy machine, the option of noise control at the source as well as designing for quietness is not available. The next option is to control the noise in the path making use of acoustic barrier, hood, enclosure, etc. for the casing radiated noise, muffler or silencer for the duct-borne intake/exhaust noise, vibration isolator to reduce propagation of the unbalanced forces to the foundation or the support structure, and structural discontinuities (impedance mismatch) to block propagation of the structure-borne sound. All these measures have been discussed at length in Chapters 4, 6 and 7. As indicated there, damper plays the same role in absorption of structure-borne sound as acoustically absorptive material does in dissipation of airborne sound. This is illustrated hereunder.

	Reflective measures	Absorptive measures
Airborne sound	Sound barrier	Sound absorber
Structure-borne sound	Vibration isolator	Vibration damper

In practice, however, noise control measures are not purely reflective (reactive) or purely absorptive (dissipative). For example, a stand-alone acoustic enclosure consists of both types of components — impervious layer for reflection and acoustic lining on the inside for absorption. Similarly, viscoelastic pads combine elasticity and damping.

Finally, it may be noted that in practice, despite the cost disadvantage, noise is controlled more often in the path than at the source, because of logistic convenience and ready availability of acoustic enclosures, mufflers or silencers, vibration isolators and dampers.

8.4 Noise Control at the Receiver End

This option is the last resort, as it were. The receiver may be protected from excessive noise exposure by means of:

(a) ear plugs or muffs (which are really a path control measure),
(b) rotation of duties (part-time removal from the noisy environment),
(c) cabin for the operator, driver or foreman (which is again a path control measure),
(d) control room for the supervisory personnel in noisy test cells.

Ear plugs and muffs are readily available in the market. They provide an insertion loss of 5–15 dB for the user, but then the user's functionality may be compromised. He/she may not be able to detect a malfunction in the machine or may not be forewarned of the damages arising from a malfunction in a machine in the vicinity.

Design of acoustic cabins and control rooms is essentially similar to the design of acoustic enclosures discussed in Chapter 6.

It may be noted that noise control at the receiver end may protect a particular receiver from the noise but would not help others in the vicinity. This comment would also apply to some of the path control measures. Therefore, noise control at the source or designing for quietness is the most cost-effective way that would help everybody in the vicinity, not just the operator or driver of the machine.

8.5 Noise Control of an Existing Facility

If there is a complaint about environmental noise due to an existing facility like captive power station, compressed air facility, an HVAC system, or a workshop, we may proceed as follows:

(a) Make SPL measurements so as to identify and rank order different sources of noise.
(b) Calculate the extent of noise reduction required for the noisiest of the sources (machines or processes) in order to reduce the total noise in the neighborhood to the desired level.

(c) Identify the paths of noise transmission — structure borne as well as airborne — from the major sources to the neighborhood and plan the path control measures.

(d) Estimate the cost-effectiveness of different alternatives.

(e) Carry out detailed engineering and installation of the selected noise control measures.

(f) Verify the effectiveness of the implemented measures and work out refinements or corrections as necessary to meet the environmental noise limits.

For community noise, it is advisable that at worst, any facility should not increase background (or ambient) noise levels in a community by more than 5 dB (A) over existing levels without the facility, irrespective of what regulations may allow.

If complaints arise from the workplace, then regulations should be satisfied, but to minimize hearing damage compensation claims, the goal of any noise control program should be to reach a level of no more than 85 dB (A).

It is important to point out that we should not wait for complaints to arise. Once a person files an official complaint, he/she would not be satisfied by normal noise reduction. A better or more desirable way is to carry out an environmental impact assessment (EIA) before a potentially noisy facility is installed. If the projected noise levels at the property line exceed the mandated limits (depending on the category of the locality or area), then either quieter machines may be ordered/specified, or path control measures may be specified for the relatively noisy machines. This is discussed in the following sections.

The EIA exercise needs prior knowledge of the total sound power level, the spectral content, and the relative locations of all the noisy machines and processes that would constitute the proposed facility. The next few sections list empirical expressions for sound power levels (or sound pressure level at 1 m distance) of some of the most common (and noisiest) machines and processes, gleaned from the literature [1,2]. These expressions, incidentally, give us a glimpse of the parameters and considerations for selection or design of quieter machines and processes. This is in fact the primary purpose of listing these expressions here in this chapter dealing with noise control strategies.

8.6 Estimation and Control of Compressor Noise

Sound power level generated within the exit piping of large centrifugal compressors (> 75 kW) is given by [1, 3].

$$L_w = 20 \log kW + 50 \; \log U - 46 \; \text{dB} \tag{8.1}$$

where U is the impeller tip speed, (m/s), $30 < U < 230$ and kW is the power of the driver motor in kilowatts.

The frequency of maximum (peak) noise level is given by

$$f_p = 4.1 \; U \; (\text{Hz}) \tag{8.2}$$

The sound power level in the octave band containing f_p is taken as 4.5 dB less than the overall sound power level. The octave-band frequency spectrum rolls off at the rate of 3 dB per octave above and below the band of maximum noise level.

Equation (8.1) indicates that sound power level of large centrifugal compressor increases with second power of the driver motor power (power index = 2) and fifth power (power index = 5) of the impeller tip speed. Thus, L_w is a much stronger function of the impeller tip speed, U. It is obvious that if we want to reduce the compressor noise, we must design it for the lowest tip speed:

$$U = \pi D.\text{RPM}/60 \; (\text{m/s}) \tag{8.3}$$

where D is diameter of the impeller, and RPM denotes the impeller rotational speed in revolutions per minute.

Example 8.1. Evaluate the overall A-weighted sound power level of a centrifugal compressor of 100 kW power with impeller diameter of 0.9 m turning at 3000 RPM.

Solution. Use of Eq. (8.3) yields the impeller tip speed, U:

$$U = \frac{\pi * 0.9 * 3000}{60} = 141.4 \; \text{m/s}$$

Use of Eq. (8.1) yields

$$L_w = 20 \log 100 + 50 \log 141.4 - 46 = 101.5 \text{ dB}$$

Equation (8.2) gives the peak frequency, f_n:

$$f_p = 4.1 \times 141.4 = 580 \text{ Hz}$$

Obviously, f_p falls in the 500-Hz octave band. In this band, the sound power level would be $101.5 - 4.5 = 97$ dB, and in the neighboring bands it will fall at the rate of 3 dB per octave. The rest of the calculations are shown in the following table:

Octave band mid-frequency (Hz)	63	125	250	500	1000	2000	4000	8000
Band sound power level	88	91	94	97	94	91	88	85
A-weighting correction (from Table 1.2)	−28.2	−18.1	−8.6	−3.2	0	1.2	1.0	−1.1
A-weighted power level	61.8	74.9	85.4	93.8	94	92.2	89.0	83.9

Finally, the overall A-weighted sound power level is evaluated by means of Eq. (1.37), making use of the last row of the table above. Thus, overall A-weighted sound power level of the compressor is given by

$$L_{wA} = 10 \log(10^{6.18} + 10^{7.49} + 10^{8.54} + 10^{9.38} + 10^{9.4} + 10^{9.22} + 10^{8.9} + 10^{8.39})$$

$$= 99.0 \text{ dB}$$

For comparison, the overall sound power level within the exit piping of a reciprocating compressor is given by

$$L_W = 106.5 + 10 \log(kW) \tag{8.4}$$

Its octave-band frequency spectrum may be calculated using the same procedure as for the centrifugal compressor above, except that the peak frequency for a reciprocating compressor is given by

$$f_p = B \times RPM/60 \tag{8.5}$$

where B is the number of cylinders of the compressor. Interestingly, a 100-kW reciprocating compressor would generate a sound power level of 128.5 dB as per Eq. (8.4), which is 25 dB more than the corresponding centrifugal compressor of Example 8.1.

Exterior noise levels of large compressors are given by the following empirical expression [4]:

Centrifugal Compressors:

$$L_W \text{ (casing)} = 79 + 10 \, \log(kW) \tag{8.6}$$

$$L_W \text{ (inlet)} = 80 + 10 \, \log(kW) \tag{8.7}$$

Rotary and reciprocating compressors (including partially muffled inlets):

$$L_W \text{ (casing)} = 90 + 10 \, \log(kW) \tag{8.8}$$

Equations (8.6)–(8.8) indicate clearly that in general rotary and reciprocating compressors are noisier than the corresponding centrifugal compressors by $90 - (79 \oplus 80) = 90 - 82.5 = 7.5$ dB. Therefore, for quieter installations, we must prefer centrifugal compressors to the rotary and reciprocating compressors.

The corresponding overall A-weighted sound power levels may be evaluated by making use of Table 8.1 and the procedure outlined above in Example 8.1. The values listed in Table 8.1 are to be subtracted from the overall sound power level.

Incidentally, compressors with lower power ($<$75 kW) are not much quieter than those with larger power. An estimate of sound pressure level at 1 m distance from the body of the compressor may be had from table 8.2, which lists the SPL values measured in early 1980s, i.e., about four decades ago. Noise control technologies have evolved a lot since then and the SPL values of the present-day compressors are known to be lower by several decibels [1]. In other words, the values listed in Table 8.2 are conservative (on the higher side) by 5–10 dB.

It may be noted that SPL of compressors is a weak function of the air compressor power, kW.

Table 8.1. Octave band correction for exterior noise levels radiated by compressors (adapted from Ref. [4]).

Octave band center frequency (Hz)	Correction (dB)		
	Rotary and reciprocating	Centrifugal, casing	Centrifugal, air inlet
31.5	11	10	18
63	15	10	16
125	10	11	14
250	11	13	10
500	13	13	8
1000	10	11	6
2000	5	7	5
4000	8	8	10
8000	15	12	16

Table 8.2. Estimated sound pressure levels of small air compressors at 1 m distance in dB (adapted from Ref. [2]).

Octave band center frequency (Hz)	Air compressor power (kW)		
	Up to 1.5	2–6	7–75
31.5	82	87	92
63	81	84	87
125	81	84	87
250	80	83	86
500	83	86	89
1000	86	89	92
2000	86	89	92
4000	84	87	90
8000	81	84	87

The overall sound power level of a compressor may be evaluated from the overall sound pressure level listed in Table 8.2 by means of the survey method:

$$SWL \equiv L_w = L_p\,(1m) + 10\,\log S_m \qquad (8.9)$$

where S_m is area of the hypothetical rectangular surface surrounding the compressor, at 1 m distance from the nominal compressor body.

Main sources of noise of small reciprocating hermetically sealed piston compressors are [5]:

(a) gas flow pulsation through the inlet and discharge valves and pipes,
(b) gas flow fluctuations in the shell cavity, which excite the cavity and shell modes,
(c) vibrations caused by the mechanical system rotation of the drive shaft and out-of-balance reciprocating motion of the piston and connecting rod, and
(d) electric motor noise.

All these mechanisms contribute directly or indirectly to the compressor shell vibration response and result in shell sound radiation. Therefore, noise control measures for these refrigerator compressors include:

(i) improved design of suction muffler,
(ii) vibro-acoustic design of the shell making use of FEM and/or BEM.

For medium and large compressors, noise control is normally done by means of intake mufflers and acoustic wrapping, hood or a stand-alone enclosure as discussed later in Section 8.11.1 on turbine noise control. Designing the compressors of different types for quietness calls for a detailed knowledge of the working of the compressor, its constituent components, and the interaction thereof [5].

8.7 Estimation and Control of Noise of Fans and Blowers

Sound power of axial fans and blowers (centrifugal fans) may be described in terms of the Specific Sound Power, H, defined by

Madison [6] as follows:

$$H = \frac{W}{(\Delta P)^2 Q} \Rightarrow L_w = L_H + 10 \log Q + 20 \log(\Delta P) \qquad (8.10)$$

where ΔP is the static pressure rise, or the stagnation pressure drop or back pressure on the fan or blower, in kilopascals (kPa), and Q is the volumetric flow rate in m^3/s. Based on the research of Graham and Hoover [7] and others, Table 8.3 gives specific sound power levels (L_H) in the eight primary octave bands for different types of fans and blowers [8]. In the logarithmic form of Eq. (8.10), ΔP is in kPa and the constant term arising from logarithms has been absorbed in L_H.

Table 8.3. Specific power levels in eight lowest octave bands for a variety of axial and centrifugal fans (adapted with permission from Ref. [8]).

Fan type	Rotor diameter (m)	63	125	250	500	1k	2k	4k	8k	Add for BPF
Backward curved	>0.75	85	85	84	79	75	68	64	62	3
centrifugal	<0.75	90	90	88	84	79	73	69	64	3
Forward curved	All	98	98	88	81	81	76	71	66	2
centrifugal										
Low-pressure radial	>1.0	101	92	88	84	82	77	74	71	7
$996 \leq \Delta P \leq 2490$	<1.0	112	104	98	88	87	84	79	76	7
Mid-pressure radial	>1.0	103	99	90	87	83	78	74	71	8
$2490 \leq \Delta P \leq 4982$	<1.0	113	108	96	93	91	86	82	79	8
High-pressure radial	>1.0	106	103	98	93	91	89	86	83	8
$4982 \leq \Delta P \leq 14945$	<1.0	116	112	104	99	99	97	94	91	8
Vaneaxial										
$0.3 \leq D_h/D \leq 0.4$	All	94	88	88	93	92	90	83	79	6
$0.4 \leq D_h/D \leq 0.6$	All	94	88	91	88	86	81	75	73	6
$0.6 \leq D_h/D \leq 0.8$	All	98	97	96	96	94	92	88	85	6
Tubeaxial	>1.0	96	91	92	94	92	91	84	82	7
	<1.0	93	92	94	98	97	96	88	85	7
Propeller	All	93	96	103	101	100	97	91	87	5

Here, BPF is the blade passing frequency, and the last column in Table 8.3 gives the value to be added to the level of the particular band in which the BPF falls.

Example 8.2. Estimate the A-weighted sound power level of a 16-bladed backward-curved centrifugal fan running at 3000 RPM and delivering 50,000 m³/h of air against back pressure of 5000 Pa. The rotor tip diameter is 0.9 m.

Solution.

Flow rate $Q = \dfrac{50000}{3600} = 13.89$ m³/s

Back pressure, $\Delta P = \dfrac{5000}{1000} = 5\,\text{kPa}$

$$10 \log Q + 20 \log(\Delta P) = 10 \log 13.89 + 20 \log 5$$
$$= 11.4 + 14.0 = 25.4 \text{ dB}$$

$BPF = \frac{16*3000}{60} = 800$ Hz, and this falls in the 1000-Hz octave band.

So, referring to the first row of Table 8.3 we must add 3 dB to the L_H value of 75 dB at the 1000 Hz octave band. Thus, we can construct the following table:

Octave band center frequency (Hz)	63	125	250	500	1000	2000	4000	8000
Specific sound power level, L_H (dB)	85	85	84	79	78	68	64	62
L_W (band) = $L_H + 25.4$ (dB)	110.4	110.4	109.4	104.4	103.4	93.4	89.4	87.4
A-weighting correction	−26.2	−16.1	−8.6	−3.2	0	1.2	1.0	−1.1
L_{WA} (octave) (dB)	84.2	94.3	100.8	101.2	103.4	94.6	90.4	86.3

Finally, overall L_{WA} may be obtained by logarithmically adding the octave band values of L_{WA} from the last row. Thus, overall A-weighted SWL is given by

$$L_{WA} = 10 \log \left(\frac{10^{8.42} + 10^{9.43} + 10^{10.08} + 10^{10.12}}{+10^{10.34} + 10^{9.46} + 10^{9.04} + 10^{8.63}} \right) = 107.4 \text{ dB}$$

Design of quieter fans (for automotive engine, for example) would normally involve:

(i) reducing the fan impeller tip speed by reducing the RPM and/or impeller diameter,

(ii) increasing the number of blades so as to reduce the required RPM for a given flow rate and static pressure rise,

(iii) using thermal (temperature controlled) drive for the automobile fans so that the fan would switch off automatically at higher automobile speeds,

(iv) using airfoil blades in order to increase the aerodynamic efficiency,

(v) using non-metallic blades (for increased loss factor),

(vi) increasing the cooling efficiency so that a fan with lower tip speed would suffice,

(vii) minimizing restriction to the airflow so that the fan could work under lower back pressure,

(viii) increasing the radiator frontal area and improving the radiator design in order to reduce the cooling load on the fan, and

(ix) using a shroud with minimal radial blade tip clearance in order to reduce the recirculation of flow around the tip.

For a given fan installation, noise can be reduced by means of:

(a) intake silencers,

(b) discharge or exhaust silencers,

(c) acoustic wrapping, hood or enclosure around the fan in order to contain and absorb the casing noise (discussed in Chapter 6).

A stand-alone acoustic enclosure would incorporate all three of these measures. The intake and exhaust silencers would then take

Table 8.4. Estimated sound pressure levels of packaged chillers at one meter (adapted from Ref. [2]).

Type and cooling capacity of machine	Octave band center frequency (Hz)									A-weighted (dBA)
	31.5	63	125	250	500	1k	2k	4k	8k	
Reciprocating compressors										
10–50 tons	79	83	84	85	86	84	82	78	72	89
51–200 tons	81	86	87	90	91	90	87	83	78	94
Rotary screw compressors										
100–300 tons	70	76	80	92	89	85	80	75	73	90
Centrifugal compressors										
Under 500 tons	92	93	94	95	91	91	91	87	80	97
500 tons or more	92	93	94	95	93	98	98	93	87	103

the shape of parallel baffle mufflers or acoustic louvres (discussed in Chapter 7).

8.8 Estimation and Control of Noise of Packaged Chillers

Compressor is the main source of noise in packaged chillers. Measured values of SPL at 1 m from the body of the chiller (or, for that matter, the compressor within) are listed in Table 8.4. These levels are generally higher than observed [1]. For more reliable values, we should refer to the manufacturer's data, if provided.

The octave band levels given in Table 8.4 may be first corrected for A-weighting and then added logarithmically to estimate the overall A-weighted sound pressure level of the packaged chiller. The chiller noise may be controlled by means of acoustic wrapping, hood or a stand-alone acoustic enclosure.

8.9 Estimation and Control of Noise of Cooling Towers

Fan or blower is the main source of noise in the cooling towers, also known as remote radiators. Their sound power level, therefore, depends on the fan or blower used therein, and is given by the following empirical expressions [1, 2]:

Propeller-type cooling towers:

Fan power up to 75 kW: $L_W = 100 + 8\log(kW)$ $\hspace{2cm}$ (8.11)

Fan power greater than 75 kW: $L_W = 96 + 10\log(kW)$ $\hspace{1cm}$ (8.12)

Centrifugal type cooling towers:

Fan power up to 60 kW: $L_W = 85 + 11\log(kW)$ $\hspace{2cm}$ (8.13)

Fan power greater than 60 kW: $L_W = 93 + 7\log(kW)$ $\hspace{1cm}$ (8.14)

The octave band sound power levels in different octave bands may be calculated by subtracting the values listed in Table 8.5, and finally, the overall A-weighted SWL can be calculated as per Example 8.2.

Equations (8.11)–(8.14) are functions of power only. It is understood that the fan or blower is running at its rated speed (RPM). If the speed is half of the rated speed, then SWL would be lower by about 8 dB [1]. Table 8.5 may be used to evaluate the sound power level in different octave bands. Finally, the overall A-weighted sound power level may be calculated as illustrated in Example 8.2 in Section 8.7.

Table 8.5. Values (dB) to be subtracted from the overall SWL of cooling towers to obtain the octave band SWL (adapted from Ref. [2]).

Octave band center frequency (Hz)	Propeller type	Centrifugal type
31.5	8	6
63	5	6
125	5	8
250	8	10
500	11	11
1000	15	13
2000	18	12
4000	21	18
8000	29	25
A-weighted (dBA)	9	7

Noise of the cooling tower can be reduced by enclosing it with an acoustic enclosure with the two longer opposite sides made entirely of acoustic louvers with sufficient air passages to provide adequate air for the cooling operation.

8.10 Estimation and Control of Pump Noise

Hydraulic pumps are used in processing plants, climate control systems, hydraulic fluid power, clean water supply, etc. Empirical expressions of the overall SPL of pumps at 1 m distance are listed in Table 8.6 for different speed ranges, as a function of the nominal drive motor power.

The octave-band frequency adjustments for the pump SPL are listed below in Table 8.7.

The overall A-weighted SPL at 1 m may be calculated as explained in Example 8.2 and SWL may be calculated by means of Eq. (8.9). Overall A-weighted sound power level of pumps of different types may directly be calculated by means of Table 8.8 (adapted from Cudina [9]). Operation of a pump creates pressure pulsation which can be spread by pipes as structure-borne noise and by the liquid medium as fluid-borne noise around the whole pumping system.

Table 8.6.   Overall sound pressure levels at 1 m from the pump (adapted from Ref. [2]).[a]

Speed range (rpm)	Nominal drive motor power	
	Under 75 kW	Above 75 kW
3000–3600	72 + 10 log (kW)	86 + 3 log (kW)
1600–1800	75 + 10 log (kW)	89 + 3 log (kW)
1000–1500	70 + 10 log (kW)	84 + 3 log (kW)
450–900	68 + 10 log (kW)	82 + 3 log (kW)

Note: [a]Subtract 2 dB to obtain the corresponding A-weighted values.

Table 8.7. Frequency adjustments for pump sound pressure levels (adapted from Ref. [2]).

Octave band center frequency (Hz)	Values to be subtracted from overall sound pressure level (dB)
31.5	13
63	12
125	11
250	9
500	9
1000	6
2000	9
4000	13
8000	19

Table 8.8. Prediction of the A-weighted sound power level generated by different pumps (adapted with permission from Ref. [9]).

Type of pump	A-weighted sound power level, (P in kW, $P_{ref} = 1$ kW) in dB	Valid for power consumption P
Centrifugal pumps (single stage)	$L_{WA} = 71 + 13.5 \log P \pm 7.5$	$4 \text{ kW} \leq P \leq 2000 \text{ kW}$
Centrifugal pumps (multistage)	$L_{WA} = 83.5 + 8.5 \log P \pm 7.5$	$4 \text{ kW} \leq P \leq 20,000 \text{ kW}$
Axial-flow pumps	$L_{WA} = 78.5 + 10 \log P \pm 10$ at Q_{BEP} $L_{WA} = 21.5 + 10 \log P$ $+ 57 Q/Q_{BEP} \pm 8$	$10 \text{ kW} \leq P \leq 1300 \text{ kW}$ $0.77 \leq Q/Q_{BEP} \leq 1.25$
Multi-piston pumps (inline)	$L_{WA} = 78 + 10 \log P \pm 6$	$1 \text{ kW} \leq P \leq 1000 \text{ kW}$
Diaphragm pumps	$L_{WA} = 78 + 9 \log P \pm 6$	$1 \text{ kW} \leq P \leq 100 \text{ kW}$
Screw pumps	$L_{WA} = 78 + 11 \log P \pm 6$	$1 \text{ kW} \leq P \leq 100 \text{ kW}$
Gear pumps	$L_{WA} = 78 + 11 \log P \pm 3$	$1 \text{ kW} \leq P \leq 100 \text{ kW}$
Lobe pumps	$L_{WA} = 84 + 11 \log P \pm 5$	$1 \text{ kW} \leq P \leq 10 \text{ kW}$

Figure 8.1 shows some devices for reduction of pressure pulsation in piping (structure-borne noise), and Fig. 8.2 illustrates some measures for reduction of structural vibration [9].

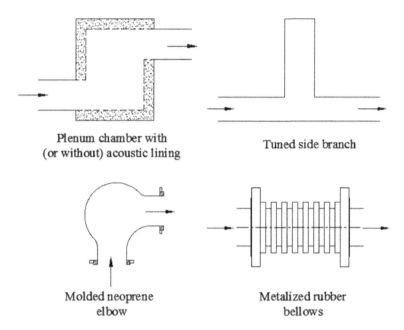

Fig. 8.1   Schematics of devices for reduction of pressure pulsation in piping [9].

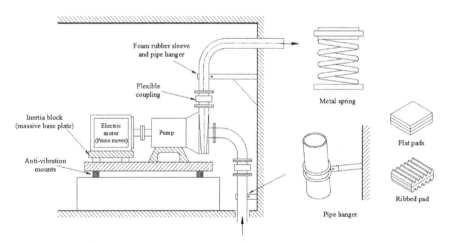

Fig. 8.2   Schematic of different measures for reduction of structure-borne sound [9].

8.11 Estimation and Control of Noise of Prime Movers

Conventional prime movers are turbines, reciprocating piston engines (particularly diesel engines and gas-driven engines), and electric motors. Empirical expressions and the associated octave-band frequency corrections, all adopted from Refs. [1, 2] are listed below. As indicated earlier, the primary source of data [2] being nearly three decades old, the estimates are considerably on the higher side; the present-day prime movers are designed, fabricated and maintained to be relatively quieter by several decibels [1].

8.11.1 *Turbines*

Sources of noise in a typical turbine are the casing, inlet and exhaust. The overall sound power level of these sources for a gas turbine are given by the following empirical expressions [2], followed by some comments on the control measures:

Casing:

$$L_W = 120 + 5\log (\text{MW}) \tag{8.15}$$

Inlet:

$$L_W = 127 + 5\log (\text{MW}) \tag{8.16}$$

Exhaust:

$$L_W = 133 + 10\log (\text{MW}) \tag{8.17}$$

Incidentally, steam turbines are considerably quieter than the gas turbines, as may be observed from the following empirical expressions [4]:

$$L_W = 93 + 4\log(kW) = 105 + 4\log (\text{MW}) \tag{8.18}$$

Frequency corrections for different octave bands for the three noise sources of gas turbine and the overall sound power level of steam turbine are listed in Table 8.9.

It may be observed from Eqs. (8.15)–(8.17) that exhaust noise is the largest source of turbine noise, followed by the intake noise and the casing noise in that order. The duct-borne exhaust noise and

Table 8.9. Frequency adjustments (in dB) for gas turbine and steam turbine noise levels (adapted from Refs. [1, 2]).

Octave band center frequency (Hz)	Value to be subtracted from overall L_W (dB)			
	Gas turbine			
	Casing	Inlet	Exhaust	Steam turbine
31.5	10	19	12	11
63	7	18	8	7
125	5	17	6	6
250	4	17	6	9
500	4	14	7	10
1000	4	8	9	10
2000	4	3	11	12
4000	4	3	15	13
8000	4	6	21	17
A-weighted (dB(A))	2	0	4	5

Note: Subtract these values from the overall sound power level L_W, to obtain octave band and A-weighted sound power levels.

intake noise are absorbed by means of dissipative parallel baffle mufflers designed as per Chapter 7, and the casing noise is contained by means of acoustic lagging, hood or enclosure designed as per Chapter 6. Table 8.10 lists the noise reduction or insertion loss in the turbine casing noise due to different noise control treatments [1, 2].

Unfortunately, however, Table 8.10 does not give the overall reduction in the A-weighted sound power Level due to each of the five treatments. This may be calculated as illustrated in Example 8.3.

Example 8.3. Calculate the A-weighting corrections for the casing noise of a 2-MW gas turbine for each of the five acoustic treatments listed in Table 8.10.

Solution. Using Eq. (8.15) we get the overall sound power level generated by the casing of the 2-MW gas turbine as follows:

$$L_W = 120 + 5\log(2) = 122\,\text{dB}$$

Table 8.10. Approximate noise reduction of gas turbine casing enclosures (adapted from Refs. [1, 2]).

Octave band center frequency (Hz)	Type 1[a]	Type 2[b]	Type 3[c]	Type 4[d]	Type 5[e]
31.5	2	4	1	3	6
63	2	5	1	4	7
125	2	5	1	4	8
250	3	6	2	5	9
500	3	6	2	6	10
1000	3	7	2	7	11
2000	4	8	2	8	12
4000	5	9	3	8	13
8000	6	10	3	8	14

Notes: [a]Glass fiber or mineral wool thermal insulation with lightweight foil cover over the insulation.
[b]Glass fiber or mineral wool thermal insulation covered with a minimum 20-gauge aluminum or 24-gauge steel.
[c]Enclosing metal cabinet for the entire packaged assembly, with open ventilation holes and with no acoustic absorptive lining inside the cabinet.
[d]Enclosing metal cabinet for the entire packaged assembly, with open ventilation holes and with acoustic absorptive lining inside the cabinet.
[e]Enclosing metal cabinet for the entire packaged assembly with all ventilation holes into the cabinet muffled and with acoustic absorptive lining inside the cabinet.

Making use of the casing column of Table 8.9 and the octave-band noise reduction (NR) due to the five types of treatment from Table 8.10 helps us to construct the following table:

Octave band mid-frequency (Hz)		63	125	250	500	1k	2k	4k	8k
(a)	Value to be subtracted from L_W (dB)	7	5	4	4	4	4	4	4
(b)	Octave-band L_W $= 121.5-$ (a) (dB)	115	117	118	118	118	118	118	118
(c)	A-weighted correction (dB)	-26	-16	-9	-3	0	1	1	-1
(d)	Octave-band $L_{WA} =$ (b)+(c) (dB)	88	100	109	114	118	119	119	116
(e)	NR due to treatment type 1 (dB)	2	2	3	3	3	4	5	6

Octave band mid-frequency (Hz)		63	125	250	500	1k	2k	4k	8k
(f)	NR due to treatment type 2 (dB)	5	5	6	6	7	8	9	10
(g)	NR due to treatment type 3 (dB)	1	1	2	2	2	2	3	3
(h)	NR due to treatment type 4 (dB)	4	4	5	6	7	8	8	8
(i)	NR due to treatment type 5 (dB)	7	8	9	10	11	12	13	14
(j)	L_{WA} with treatment type 1 = (d)–(e) (dB)	86	98	106	111	115	115	114	110
(k)	L_{WA} with treatment type 2 = (d)–(f) (dB)	83	95	103	108	111	111	110	106
(l)	L_{WA} with treatment type 3 = (d)–(g) (dB)	87	99	107	112	116	117	116	113
(m)	L_{WA} with treatment type 4 = (d)–(h) (dB)	84	96	104	108	111	111	111	108
(n)	L_{WA} with treatment type 5 = (d)–(i) (dB)	81	92	100	104	107	107	106	102

Logarithmic addition of the octave band values by means of Eq. (1.37) yields:

row (b): L_W without any treatment (bare casing) = 126.1 dB
row (d): L_{WA} without any treatment (bare casing) = 124.5 dB

The overall A-weighted correction for the bare casing = 124.5 – 126.1 = −1.6 dB.

row (j): L_{WA} with treatment type 1 = 120.4 dB ⇒ overall NR = 124.5 – 120.4 = 4.1 dBA
row (k): L_{WA} with treatment type 2 = 116.6 dB ⇒ overall NR = 124.5 – 116.6 = 7.9 dBA
row (l): L_{WA} with treatment type 3 = 122.1 dB ⇒ overall NR = 124.5 – 122.1 = 2.4 dBA
row (m): L_{WA} with treatment type 4 = 117.1 dB ⇒ overall NR = 124.5 – 117.1 = 7.4 dBA
row (n): L_{WA} with treatment type 5 = 112.6 dB ⇒ overall NR = 124.5 – 112.8. = 11.9 dBA

The following observations may be made from the results of Example 8.3:

(i) For the noise radiated by the casing of a gas turbine, overall A-weighting correction is −1.6 dB. When rounded up to integer value, this tallies with the value of −2 dB (2 dB to be subtracted from the total linear sound power level) given in the last row of Table 8.9.

(ii) Simply wrapping the turbine (or for that matter, any noisy machine) with a blanket of absorptive material gives a noise reduction of only 4.1 dBA (see row j in the example above); thin protective foil (of about 40 micron or 0.04 mm thickness) is of no consequence, acoustically speaking.

(iii) Replacing the thin foil (of negligible surface density) with 24-gauge steel (thickness about 0.8 mm, and surface density of $7800 * 0.8/1000 = 6.24$ kg/m^2) helps considerably in that the noise reduction increases from 4.1 to 7.9 dBA.

(iv) Enclosing the turbine with a metal cabinet without any acoustic lining and keeping the ventilation holes open is more or less useless in that it reduces noise by 2.4 dBA only. This follows from the relationship between insertion loss (IL) and transmission loss (TL), as explained in Chapter 6.

(v) Lining the metal cabinet with an acoustical blanket on the inside, while keeping the ventilation holes open, restores the insertion loss or noise reduction to 7.4 dBA, as for the type 2 treatment.

(vi) Finally, a proper acoustical cabinet with all ventilation holes or openings muffled yields a noise reduction of 11.9 dBA, or about 12 dBA. However, this is much less than the insertion loss of 20 dBA that can be obtained by means of a properly designed, fabricated and installed stand-alone (or walk-around) acoustical enclosure, as described in Chapter 6. This brings out the inherent limitations of an acoustic hood or cabinet because of the dynamic or inertial coupling of the thin layer of air in between the machine (turbine, in this case) and tight-fitting hood or cabinet.

8.11.2 *Diesel engines*

Diesel engines are used widely in automobiles as well as captive diesel generator (DG) sets. Empirical expression for the sound power level of the diesel engine exhaust noise is given by [1, 2]

$$L_W = 120 + 10\log(kW) - K - (l_{ex}/1.2) \qquad (8.19)$$

where $K = 0$ for a naturally aspirated engine, and $K = 6$ for a turbocharged engine; kW denotes the nominal rated power of the engine in kilowatts; and l_{ex} is length of the long exhaust pipe (or tailpipe) in meters. Equation (8.19) implies that attenuation due to wall friction and eddy losses is 0.83 dB per meter length of the exhaust pipe.

The octave band frequency spectrum of exhaust noise may be obtained by subtracting the values listed in Table 8.11 from SWL of Eq. (8.19).

Sound power level of the turbocharged diesel engine inlet noise is given by the empirical expression [2]

$$L_W = 95 + 5\log(kW) - (l_{in}/1.8) \qquad (8.20)$$

Table 8.11. Frequency adjustments for unmuffled engine exhaust noise and A-weighted level (adapted from Ref. [2]).

Octave band center frequency (Hz)	Value to be subtracted from overall sound power level (dB)
31.5	5
63	9
125	3
250	7
500	15
1000	19
2000	25
4000	35
8000	43
A-weighted (dB(A))	12

Table 8.12. Frequency adjustments (dB) for turbo-charged air inlet (adapted from Ref. [2]).

Octave band center frequency (Hz)	Value to be subtracted from overall sound power level (dB)
31.5	4
63	11
125	13
250	13
500	12
1000	9
2000	8
4000	9
8000	17
A-weighted (dB(A))	3

Note: Subtract these values from the overall sound power level (Eq. (8.20)) to obtain octave band and A-weighted level.

where l_{in} is length of the inlet or induction pipe. Equation (8.20) implies that attenuation of the intake noise due to wall friction and eddy losses is about 0.56 dB per meter length of the induction pipe.

The octave band frequency spectrum may be obtained by means of Eq. (8.20) and Table 8.12.

Sound power level generated by the diesel engine casing is given by the empirical expression [1, 2]:

$$L_W = 94 + 10\log(kW) + A + B + C + D \qquad (8.21)$$

Constants A, B, C and D are listed below in Table 8.13. The octave band frequency spectrum of the diesel engine casing noise may be evaluated by subtracting the corrections of Table 8.14 from Eq. (8.21).

8.11.3 *Electric motors*

A rough (and conservative) estimate of the sound pressure level (at 1 m) of the totally enclosed, fan cooled (TEFC) small electric motors is given by [1, 2]:

Table 8.13. Correction terms to be applied to Eq. (8.21) for estimating the overall sound power level of the casing noise of a reciprocating engine (adapted from Ref. [2]).

Speed correction term, A	
Under 600 rpm	−5
600–1500 rpm	−2
Above 1500 rpm	0
Fuel correction term, B	
Diesel only	0
Diesel and natural gas	0
Natural gas only (including a small amount of pilot oil)	−3
Cylinder arrangement, C	
In-line	0
V-type	−1
Radial	−1
Air intake correction term, D	
Unducted air inlet to unmuffled roots blower	+3
Ducted air from outside the enclosure	0
Muffled roots blower	0
All other inlets (with or without turbo-charger)	0

Table 8.14. Frequency adjustments (dB) for casing noise of reciprocating engines (adapted from Refs. [1, 2]).

Octave band center frequency (Hz)	Engine speed under 600 rpm	Engine speed 600–1500 rpm		Engine speed over 1500 rpm
		Without roots blower	With roots blower	
31.5	12	14	22	22
63	12	9	16	14
125	6	7	18	7
250	5	8	14	7
500	7	7	3	8
1000	9	7	4	6
2000	12	9	10	7
4000	18	13	15	13
8000	28	19	26	20
A-weighted (dB(A))	4	3	1	2

Note: Subtract values from the overall sound power level (Eq. (8.21)) to obtain octave band and A-weighted levels.

Table 8.15.   Octave band level adjustments (dB) for small electric motors (adapted from Ref. [2]).

Octave band center frequency (Hz)	Totally enclosed, fan cooled (TEFC) motor	Drip proof (DRPR) motor
31.5	14	9
63	14	9
125	11	7
250	9	7
500	6	6
1000	6	9
2000	7	12
4000	12	18
8000	20	27
A-weighted (dB(A))	1	4

Under 40 kW:

$$L_p(1m) = 17 + 17\log(kW) + 15\log(RPM) \qquad (8.22)$$

Over 40 kW (up to 300 kW):

$$L_p(1m) = 29 + 10\log(kW) + 15\log(RPM) \qquad (8.23)$$

The corresponding overall sound power level may be obtained by means of Eq. (8.9). Drip proof (DRPR) motors are known to be quieter than their TEFC counterparts by about 5 dB for the same power and RPM.

The octave band frequency adjustments, adapted from Refs. [1,2] are listed in Table 8.15.

For large electric motors (rated power of 750–4000 kW) with nominal acoustical hoods, one can make use of Table 8.16 for a rough estimation of the sound power level, L_w [1,2].

In order to estimate SWL of motors with rated power above 4000 kW, we should add 3 dB to all levels in Table 8.15, and for motors rated between 300 and 750 kW, we should subtract 3 dB from all sound power levels. The overall linear sound power levels may be obtained by means of logarithmic addition. Similarly, the overall A-weighted sound power levels may be obtained by first

Table 8.16. Sound power levels of large electric motors (adapted from Refs. [1, 2]).

Octave band center frequency (Hz)	1800 and 3600 rpm	1200 rpm	900 rpm	720 rpm and lower	250 and 400 rpm vertical
31.5	94	88	88	88	86
63	96	90	93	90	87
125	98	92	92	92	88
250	98	93	93	93	88
500	98	93	93	93	88
1000	98	93	96	98	98
2000	98	98	96	92	88
4000	95	88	88	83	78
8000	88	81	81	75	68

adding algebraically the A-weighting corrections, and then adding up logarithmically, as illustrated before in the solved examples 8.1–8.3.

The sources of noise and vibration in electric motors (and indeed in most electric machines including generators or alternators) are [10]:

(a) Electromagnetic forces in the airgap between the stator and rotor characterized by rotating or pulsating power waves in the range of 100–4000 Hz.

(b) Bearings, depending on the quality of manufacture, the accuracy of machining of the bearing seats and the vibroacoustic properties of the end brackets.

(c) Aerodynamic forces that depend on the construction of the fan and ventilation channels of the machine.

(d) Mechanical imbalance of rotors that may excite an appreciable amount of vibration, particularly in high-speed motors; and

(e) Rubbing of brushes against the commutator or contact rings that produces a predominately high frequency noise.

The noise control of electric motors, therefore, calls for:

(i) precision fabrication and installation of stator and rotor windings, rotors and bearings,

(ii) balancing of the rotor at site, at the time of installation, and periodic balancing thereafter, and

(iii) use of quieter (or muffled) cooling fans of TEFC motors.

8.12 Jet Noise Estimation and Control

Jet noise pre-dominates in turbojet engines, furnaces, pneumatic cleaning devices, etc. The total acoustic power (W_o) radiated by a high-velocity jet is a small, but significant, fraction of the mechanical power (W_m) of the jet. It is given by [1]

$$L_W = 10 \log W_a + 120, W_a = \eta W_m, W_m = \dot{m} U^2 / 2 = \rho U^3 \pi d^2 / 8 \tag{8.24}$$

where d is diameter (or equivalent diameter) of the jet,

$\dot{m}$ is the mass flux of the jet,

U is the mean flow velocity averaged over the cross-section of the jet,

ρ is mass density of the medium, and

η is the acoustical efficiency given by the approximate expression [3]

$$\eta \approx \left(\frac{T}{T_0}\right)^2 \left(\frac{\rho}{\rho_0}\right) K_a M^5 \tag{8.25}$$

Here, T_0 is the absolute temperature $(in \ ^\circ K)$ and ρ_0 is the mass density of the ambient medium; T and ρ are the absolute temperature and mass density of the medium of the jet; $M = U/c_0$ is Mach number relative to the ambient gas; and K_a is the acoustical power coefficient given by

$$K_a \approx 5 \times 10^{-5} \ \text{(for subsonic jets)} \tag{8.26}$$

For choked (under expanded) or supersonic jets, i.e., when pressure ratio exceeds 1.89 (for air medium), η is given by

$$\eta = \eta_{turb} + \eta_{shock} \simeq \eta_{shock} = \begin{cases} 3.16 \times 10^{-8} (PR)^{11.67} \ \text{for} \ PR \le 2.6 \\ 1.1 \times 10^{-3} (PR)^{0.745} \ \text{for} \ PR \ge 2.6 \end{cases} \tag{8.27}$$

where PR denotes the pressure ratio of the jet.

Table 8.17.   Directional correction for jets (adapted from Refs. [1,3]).

Angle from jet axis (°)	Directivity index, DI (dB)	
	Sub-sonic	Choked
0	0	− 3
20	+1	+1
40	+8	+6
60	+2	+3
80	− 4	− 1
100	− 8	− 1
120	− 11	− 4
140	− 13	− 6
160	− 15	− 8
180	− 17	− 10

It may be noted that for a pressure ratio higher than the critical pressure ratio, η may be much higher than (of the order of 100 times) the value given by Eq. (8.25). In other words, the presence of shocks in the nozzle or just outside may generate sound power levels that could be 20 (or more) dB higher than the turbulence noise expressions (8.24)–(8.26).

The jet noise is highly directional, and this directivity for choked jets is different from that of the subsonic jets, as can be seen from Table 8.17.

Finally, the frequency spectrum may be evaluated by

$$L_p(\text{f}) - L_p(\text{overall}) = -5 - 16.045 \left\{ \log \left(f/f_p \right) \right\}^2 + 2.406 \left\{ \log \left(f/f_p \right) \right\}^4 \text{ (dB)} \tag{8.28}$$

where f_p is the peak frequency given by the following approximate expression in terms of Strouhal number, N_s [3]

$$N_s = f_p d/U, N_s = 0.2 \text{ for subsonic jets} \tag{8.29}$$

This formula is an approximation inasmuch as it represents a curve that would be symmetric about $f = f_p$, but the curve in practice is not exactly symmetric [1,3].

The pronounced directivity pattern of jets is of great significance for high vertical exhaust stacks or chimneys. Then, all personnel

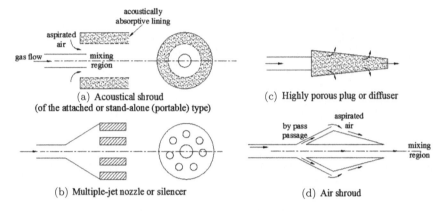

Fig. 8.3　Schematics of typical discharge silencers [1].

on the ground are at an angle of more than 90° with respect to the vertical jet axis. This is a significant noise control measure and should be duly accounted for in an EIA study.

Equations (8.24) and (8.25), when combined, indicate that the sound power generated by a fully expanded jet increases with the eighth power of the jet velocity. This underlines the necessity of reducing velocity of the jet by increasing the diameter of the jet (for the same mass flow or thrust), if we want to reduce the jet noise at the source. However, when this option is not available, or has already been utilized to the extent it was feasible, then we may make use of one of the four types of discharge silencers shown in Fig. 8.3 [1].

Noise from safety valves or relief valves with high pressure (much more than the critical pressure corresponding to sonic jet) may be reduced by means of a multi-stage trim (Fig. 8.4) or a set of properly designed multiple hole orifice plates (Fig. 8.5), adapted from Baumann and Coney [11].

Example 8.4. 30 kg/s of air is being let out to atmosphere through a 11.5 m high vertical exhaust stack of 0.5 m diameter at 25°C. Estimate the SPL of the jet noise at a location 10 m away from the base of the stack, 1.5 m above the ground.

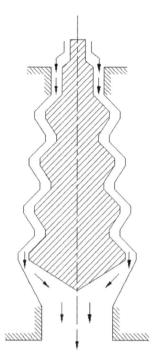

Fig. 8.4   Schematic of single flow multi-step valve plug [11].

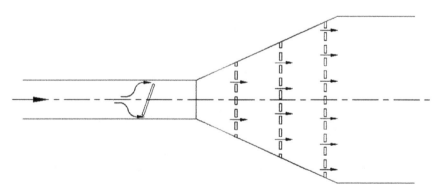

Fig. 8.5   Schematic of a butterfly valve followed by three multi-hole orifice plates for multi-stage pressure reduction [11].

Solution.

Density, $\rho_0 = \dfrac{p_0}{RT_0} = \dfrac{1.013 * 10^5}{287 * (273 + 25)} = 1.184 \text{ kg/m}^3$

Sound speed, $c_0 = (\lambda RT)^{1/2} = (1.4 * 287 * 298)^{1/2} = 346.0 \text{ m/s}$

Jet velocity, $U = \dfrac{\dot{m}}{\rho_0 A} = \dfrac{30}{1.184 \left(\dfrac{\pi}{4} * (0.5)^2 \right)} = 129 \text{ m/s}$

Mean flow Mach number, $M = \dfrac{U}{c_0} = \dfrac{129}{346} = 0.373$ (jet is subsonic)

Mechanical power of the jet,

$$W_m = \frac{\dot{m}U^2}{2} = \frac{30 * (129)^2}{2} = 2.5 * 10^5 \text{ W}$$

Acoustical efficiency, $\eta = K_a M^5 = 5*10^{-5}*(0.373)^5 = 3.61*10^{-7}$

Acoustical power, $W_a = \eta W_m = 3.61 * 10^{-7} * 2.5 \times 10^5 = 0.09$ Watt

Sound power level,

$$L_W = 10 \log W_a + 120 = 10 \log(0.09) + 120 = 109.5 \text{ dB}$$

Distance of the receiver with respect to the exhaust point of the stack is 10 m horizontal and $11.5 - 1.5 = 10$ m vertical. Thus, the receiver is located at an angle of $90 + 45 = 135°$ with respect to the jet. At this angle for a subsonic jet, referring to Table 8.17, directivity index, DI $= -12.5$ dB

$$\therefore L_W (135°) = L_W + DI = 109.5 - 12.5 = 97 \text{ dB}$$

Distance of the observer from the jet exhaust,

$$R = \left(10^2 + 10^2\right)^{1/2} = 14.14 \text{ m}$$

Finally, $L_p(R) = L_W - 10 \log \left(4\pi R^2\right)$

$$= 97 - 10 \log \left(4\pi * 14.14^2\right)$$
$$= 97 - 34 = 63 \text{ dB}$$

8.13 Estimation and Control of Gear Noise

Gears are used extensively in transmission of power in automobiles, earthmoving equipment, ship propulsion, captive power stations, home appliances, etc. Whine and rattle are two different types of gear noise. Whine is a tonal sound at the gear meshing frequency and gear rattle is an impulsive sound that occurs in lightly loaded gears excited by an externally applied oscillating torque [12].

The octave band sound pressure levels at all bands above 125 Hz, at a distance of 1 m from the spur gear system, may be estimated from the empirical expression [1, 2]

$$L_p(1 \text{ m}) = 78 + 4\log(kW) + 3\log(RPM)(\text{dB}) \qquad (8.30)$$

where RPM is the rotational speed of the slowest gear shaft.

For estimation of SPL in the 63 Hz and 31.5 Hz octave bands, one may subtract 3 and 6 dB, respectively, from the SPL given by Eq. (8.30).

As indicated earlier in this chapter, helical gears are substantially quieter than spur gears for transmission of the same power. To be more precise, the actual noise reduction (compared to a straight spur gear) is given by the approximate relationship [1]

$$\Delta SPL = 13 + 20\log Q_a \qquad (8.31)$$

where Q_a is the number of teeth that would be intersected by a straight line parallel to the gear shaft. Thus, noise reduction of up to 30 dB may be obtained by means of well-designed helical gears. Table 8.18 lists possible noise reductions consequent to change in certain design parameters [12].

It may, however, be noted that the many reductions that are possible and listed in Table 8.18 are not additive in nature.

In general, gear noise typically increases with load and power. More importantly, gear noise increases at the rate of 9–11 dB per octave beneath the torsional natural frequency [12]. Gear noise can be reduced by lowering roll angles, use of lead crown, avoiding profile errors and pressure angle errors, increasing lubricant viscosity, use of the split path drives and planetary gears, damping, use of bush

Table 8.18. Effects of different gear design and manufacturing parameters on gear noise (adapted with permission from Houser [12]).

	Direction to reduce noise	Noise reduction (dB)	Comments
Number of teeth	Decrease	0–6	Lowers mesh frequency.
Contact ratio	Increase	0–20	Requires accurate lead and profile modifications.
Helix angle	Increase	0–20	Machining errors have less effect with helical gears. Little improvement above about 35°.
Surface finish	Reduce	0–7	Depends on initial finish – reduces friction excitation.
Profile modification		4–8	Good for all types of gears.
Lapping		0–10	Very effective for hypoid gears.
Pressure angle	Reduce	0–3	Reduces tooth stiffness, reduces eccentricity effect, and increases contact ratio.
Face width	Increase		Increases contact ratio for helical gears; reduces deflections.

bearings rather than rolling element bearings, and use of housing mounts and tuned absorbers [12].

If all these design and operational changes do not produce adequate noise reduction, we may have to make use of one or all of the following noise control measures [12]:

(a) redesign of gears with higher contact ratio,
(b) adding isolators to the mounts of the gear box,
(c) providing an acoustic hood over the gear box as an extreme solution (as a last resort).

8.14 Earthmoving Equipment Noise Estimation and Control

General earthmoving and material handling machinery used in industry generally fall in one of the following groups:

(a) compaction machines like vibration rollers, vibratory plates and vibratory rammers,
(b) tracked dozers, tracked loaders and tracked excavator-loaders,
(c) wheeled dozers, wheeled loaders, wheeled excavator-loaders, dumpers, graders, loader-type landfill compactors, lift trucks, mobile cranes, nonvibrating rollers and hydraulic power packs,
(d) excavators, builders' hoists, construction winches,
(e) tower cranes, welding and power generators,
(f) compressors,
(g) lawnmowers, lawn trimmers and lawn edge trimmers.

European Union has specified limits on the permissible A-weighted sound power level of each of the seven groups of machines listed above, as a function of their net installed power (in kW), mass of the appliance (in kg), or cutting width (in cm), as applicable.

In particular, compacting machines, including vibrating rollers, and vibratory plates and rammers tend to be the noisiest. Similarly, the handheld concrete breakers and picks are also very noisy because of the vibratory action. Next in that order are tracked dozers, loaders and escalator loaders. A major source of noise in residential localities are lawn mowers and lawn trimmers. In general, wheeled dozers, loaders, dumpers, graders, landfill compactors, combustion engine driven counterbalanced lift trucks and compaction machines are marginally quieter (by a couple of decibels) than their tracked counterparts [13, 14].

A disturbing feature of the earthmoving equipment is that these machines are often used in residential localities and commercial areas. Portable noise barriers must be used to shield the immediate neighbors from the annoyingly excessive noise of these machines.

Legislative noise control has generally proved effective. The legislative limits have been progressively lowered. For example, the stage II (2008) values of L_{WA} are 2–3 dB lower than those of stage I (2002). With continued efforts of all machinery manufacturers, the present (2012) levels may be estimated to be lower than those of the stage II values by a couple of decibels.

In fact, legislative control of noise has proved to be effective for all types of industrial machines as well as automobiles and passenger aircraft in reducing their noise decade after decade. The legislative limits have been tightened accordingly, stage after stage, giving sufficient time for the designers as well as manufacturers to design for quietness making use of different practices listed earlier in this chapter.

As indicated in Section 8.6, noise of stationary compressors (driven by I.C. engines or electric motors) can be contained by means of acoustic hood or enclosure.

In most earthmoving equipment, sources of noise are:

(a) cooling fans,
(b) engines,
(c) electric motors,
(d) gear boxes (transmission systems),
(e) hydraulic pumps.

Noise control measures for all these components have been discussed earlier in this chapter. However, additional measures need to be adopted for the earthmoving equipments on the road or in residential and commercial locations. Portable acoustic barriers are used with limited but considerable success in containing the noise by several decibels at the site of operation like laying of roads, building of metro railway, elevated highways, flyovers, etc.

8.15 Impact Noise Control

Material handling operations often involve transfer of material from one level to another. A few of the typical operations are shown in Figs. 8.6 and 8.7.

Impact noise due to metallic components or parts falling into a tote box or stillage can be reduced by:

(a) reducing the height of fall by raising the stillage or tote box onto a platform, as shown in Fig. 8.6.
(b) breaking the free fall into several shorter or milder falls and dampening the impact by means of viscoelastic pads to cushion the fall, as shown in Fig. 8.7(a).

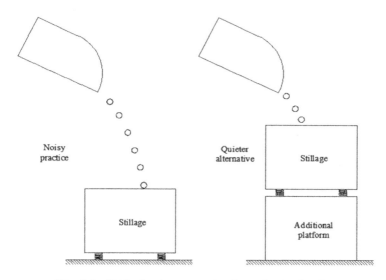

Fig. 8.6   Reducing impact noise, by reducing height.

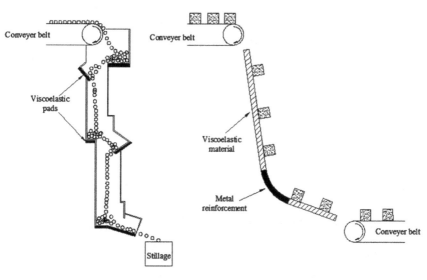

(a) Shorter and dampened falls   (b) Inclined viscoelastic chute

Fig. 8.7   Reducing impact noise, by path modification.

(c) making use of an inclined rubber chute as shown in Fig. 8.7(b); and making the stillage of Fig. 8.6 out of interwoven metallic strips rather than flat (unperforated) metallic plate.

8.16 Environmental Impact Assessment

Making use of the empirical expressions for sound power levels of various types of machines given in the last few sections of the current chapter, and the formulae presented in Chapter 6 for acoustics of rooms and enclosures, we can estimate the sound pressure level at a point of interest in the neighborhood (say, at the property line of the industry owner) with and without the noise control measures. This is illustrated in Example 8.5.

Example 8.5. A 2 MW diesel generator set is proposed to be installed at the edge of a factory at a distance of 20 m from the property line. Mandatory limits on L_{pA} for industrial area are 75 dB during the day and 70 dB during the night. Conduct an EIA to determine whether these limits would be satisfied without an acoustic enclosure. If not, what should be the minimum insertion loss for which the acoustic enclosure and exhaust muffler need to be designed. The captive power from the DG set is required during the night as well as the day. The diesel engine is turbocharged and the DG set runs at 1500 RPM. The tail pipe length is about 6 m, and induction pipe length is 0.9 m.

Solution: Use of Eqs. (8.19)–(8.21) for sound power level and the last row of Tables 8.11–8.14 for the A-weighting corrections yields rough estimates of the unmuffled sound power levels as follows:

$$L_{WA}(\text{unmuffled exhuast}) = 120 + 10\log{(2 * 1000)}$$
$$- 6 - 6/1.2 - 12 = 130 \text{ dB}$$
$$L_W(\text{inlet}) = 95 + 5\log(2 * 1000) - 0.9/1.8 - 3$$
$$= 108 \text{ dB}$$
$$L_W(\text{casing}) = 94 + 10\log(2 * 1000) - 2 = 125 \text{ dB}$$

Assuming that the engine fan noise is included in the casing noise, and that all three sources of noise are nearly equidistant (20 m) from the property line of the factory, the A-weighted SPL at the property line is given by

$$L_{pA}(r) = L_{WA} - 10\log\left(2\pi r^2\right)$$

or

$$L_{pA}(20 \text{ m}) = 130 \oplus 125 \oplus 108 - 10\log\left(2\pi * 20^2\right)$$
$$= 131 - 34 = 97 \text{ dB}$$

This SPL is 97 – 75 = 22 dB higher than the day limit, and 97 – 70 = 27 dB higher than the night-time limit.

Therefore, we need to design the acoustic enclosure and exhaust muffler for insertion loss of at least 27 dB for night-time use of the DG set. However, it is inordinately costlier to design the exhaust muffler as well as acoustic enclosure for IL exceeding 25 dB.

Significantly, however, estimates of Eqs. (8.19)–(8.21) are too conservative. As indicated in the text, the present-day machines are 5–10 dB quieter than their counterparts of four decades ago, when measurements were made for formulation of empirical formulae given in Eqs. (8.19)–(8.21). Therefore, in all probability, the exhaust muffler and acoustic enclosure designed for IL of 25 dB would suffice. Based on this EIA, installation of the DG set may be permitted subject to the factory owners installing acoustic enclosure shown in Fig. 8.8 and a properly designed residential muffler with a volume of at least 10 times the engine capacity or piston displacement (medium or large, as per Table 5.1).

The acoustic enclosure of Fig. 8.8 needs to be designed, fabricated and installed as per Section 4.3. In particular the following may be noted:

(a) For acoustic louvers with $l = 1$ m, $2d = 2h = 100$ mm, IL would be about 25 dB.
(b) The outer (impervious) layer of the acoustic enclosure may be a brick wall or at least 1.6 mm thick GI Plate.

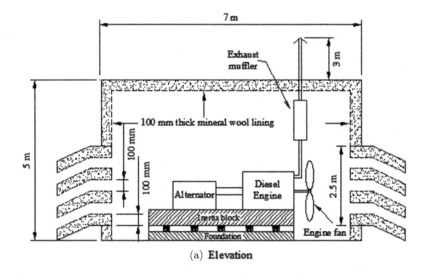

(a) **Elevation**

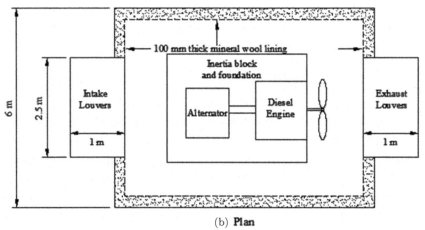

(b) **Plan**

Fig. 8.8 Schematic views of a DG set with 7 m × 6 m × 5 m (high) acoustic enclosure.

(c) For acoustic louvers and lining of the walls, the commercially available 64 kg/m^3 density mineral wool or 32 kg/m^3 density glass wool may be used. This may be covered with 8 mil (or 10 mil) glass cloth (or tissue) and a thin (26 gauge) GI plate or 22-gauge aluminum plate with at least 20% porosity.

Features 1–10 listed in Section 4.3 must be heeded. Such an acoustic enclosure would yield IL of 25 dBA at least.

References

[1] Bies, D. A. and Hansen, C. H., *Engineering Noise Control*, Fourth Edition, Spon Press, London (2009).

[2] Joint Departments of the Army, Air Force and Navy, USA, Noise and vibration control for mechanical equipment, Technical manual TM 5-805-4/AFJMAN 32-1090, Washington, DC (1995).

[3] Heitner, I., How to estimate plant noises, *Hydrocarbon Processing*, 47(2), 67–74 (1968).

[4] Edison Electric Institute, Electric power plant environmental noise guide, Bolt Beranek and Newman Inc. Technical Report (1978).

[5] Crocker, M. J., Noise control of compressors, Chapter 74, in *Handbook of Noise and Vibration Control* (Ed. Crocker, M. J.), John Wiley, New York (2007).

[6] Madison, R., *Fan Engineering (Handbook)*, Eighth Edition, Buffalo Forge Company, Buffalo, New York (1938).

[7] Graham, J. B. and Hoover, R. M., Fan noise, Chapter 41, in *Handbook of Acoustical Measurements and Noise Control*, Third Edition (Ed. Harris, C. M.), McGraw Hill, New York, (1991).

[8] Lauchle, C. C., Centrifugal and axial fan noise prediction and control, Chapter 71, in *Handbook of Noise and Vibration Control* (Ed. Crocker, M. J.), John Wiley, New York (2007).

[9] Cudina, Mirko, Pumps and pumping system noise and vibration prediction and control, Chapter 73, in *Handbook of Noise and Vibration Control* (Ed. Crocker, M. J.), John Wiley, New York (2007).

[10] Zusman, George, Types of electric motors and noise and vibration prediction and control methods, Chapter 72, in *Handbook of Noise and Vibration Control* (Ed. Crocker, M. J.), John Wiley, New York (2007).

[11] Baumann, H. D. and Coney, W. B., Noise of gas flows, Chapter 15 in *Noise and Vibration Control* (Ed. Ver, I. L. and Beranek, L. L.), Second Edition, John Wiley (2008).

[12] Houser, D. R., Gear noise and vibration prediction and control, Chapter 69, in *Handbook of Noise and Vibration Control* (Ed. Crocker, M. J.), John Wiley, New York (2007).

[13] Directive 2000/14/EC of the European Parliament and of the Council of 8 May 2000 on the Approximation of the Laws of the Member States relating to the Noise Emission in the Environment by Equipment Use Outdoors, Article 12 (2000), pp. 6–8.

[14] Bruce, R. D., Moritz, C. T. and Bommer, A. S., Sound Power Level Predictions for Industrial Machinery, Chapter 82, in *Handbook of Noise and Vibration Control* (Ed. Crocker, M. J.), Wiley, New York (2007).

Problems

8.1. For a 100-kW compressor, evaluate the total exterior sound power level generated, in dB and dBA, for the following cases:

(a) the casing of a rotary (screw type) or reciprocating compressor with a partially muffled inlet,

(b) the casing of a centrifugal compressor, and

(c) the air inlet of a centrifugal compressor,

and thence evaluate the respective A-weighting corrections ΔL_{WA} to be subtracted from the overall L_W in order to evaluate the overall L_{WA}.

[Ans.: (a) 1.0 dB, (b) 2.0 dB, and (c) 0.5 dB]

8.2. For a small compressor, making use of the estimated values of L_p (1 m), evaluate the overall SPL at 1 m without and with A-weighting corrections, and thence evaluate the overall A-weighting corrections (ΔL_{pA}) to be subtracted from the overall L_p (1 m) in order to evaluate L_{pA} (1 m) for each of the three ranges of compressor power.

[Ans.: ΔL_A = 1.1 dB for up to 1.5 kW, 1.4 dB for 2–6 kW, and 1.7 dB for 7–75 kW]

8.3. For the backward-curved centrifugal fan of Example 8.2, estimate the overall linear sound power level, and thence evaluate the overall A-weighting correction (ΔL_{WA}).

[Ans.: L_W = 115.6 dB, and ΔL_A = 8.2 dB]

8.4. For a 30-kW motor running at 1500 RPM, evaluate the overall L_{pA} at 1 m and A-weighting correction ΔL_A, for the case of a

(a) totally enclosed, fan cooled (TEFC) motor, and

(b) drip proof (DRPR) motor.

and compare the same with those given in the last row of Table 8.15.

[Ans.: L_{pA} = 89 dB, ΔL_A = 1 dB, (b) L_{pA} = 81 dB, ΔL_A = 4 dB]

8.5. A pneumatic cleaning convergent nozzle has an outlet diameter of 3 mm. The source of compressed air is a receiver with a steady static pressure of 7 bars at 25°C temperature. Estimate the sound pressure level generated by the (highly) choked jet at the operator's ear behind the nozzle at a distance of 0.5 m.

[**Ans.: 109.4 dB**]

Chapter 9

Computational Acoustics

9.1 Introduction

Computational Acoustics is a multidisciplinary field that helps in simulating and analyzing the generation, propagation, and interaction of sound waves in various environments. Computational acoustics encompasses vibro-acoustics, aeroacoustics, room acoustics, outdoor sound propagation, and underwater acoustics. However, this chapter is limited to computational methods relevant to vibro-acoustic and aeroacoustics.

Vibro-acoustic problems involve the coupling between structural vibrations and acoustic waves, and they often occur in various engineering applications such as automobiles, aerospace, industrial machinery, and architectural acoustics. Various computational methods are employed to analyse complex interactions of structural, fluid, and acoustic domains. Here are some of the computational methods used for solving vibro-acoustic problems:

- Finite Element Methods (FEMs);
- Boundary Element Methods (BEMs);
- Statistical Energy Analysis (SEA);
- Hybrid Methods;
- Wave Methods.

The choice of method depends on factors such as the problem's complexity, computational resources, accuracy requirements, and the available software tools. Often, a combination of methods may be used to accurately capture the various aspects of vibro-acoustic behavior in different frequency ranges and scenarios.

Aeroacoustics is the study of sound generation and propagation resulting from aerodynamic forces and unsteady flow. These problems often involve the complex interaction of turbulent fluid motion and sound radiation. Various methods are used to analyze and predict aerodynamic noise. Here are two of the critical methods in aeroacoustics:

- Direct Numerical Simulations (DNS);
- Acoustic analogies.

9.2 Governing Equations

The acoustic wave equation is a fundamental equation describing sound wave propagation in a gaseous medium [1]:

$$\nabla^2 p(\vec{x}, t) - \frac{1}{c^2} \frac{\partial^2 p(\vec{x}, t)}{\partial t^2} = 0 \tag{9.1}$$

where ∇^2 is the Laplacian operator (spatial derivatives), $p(\vec{x}, t)$ is the acoustic pressure as a function of spatial coordinates and time, and c is the speed of sound in the medium.

Most acoustic problems start with a linear assumption (i.e., acoustic pressure is small compared to static pressure) and simplify the problem by assuming the time-varying variables to be time-harmonic [2].

For a harmonic motion, the homogenous form of the Helmholtz equation can be written as [1]

$$\nabla^2 p(\vec{x}) + k^2 p(\vec{x}) = 0 \tag{9.2}$$

where $p(\vec{x})$ is acoustic pressure, and $k = \omega/c$ is the wave number.

The inhomogeneous form of the Helmholtz equation is [3]

$$\nabla^2 p(\vec{x}) + k^2 p(\vec{x}) = -Q\delta(\vec{x} - \vec{x}_s) \tag{9.3}$$

where Q is the source strength, $\delta(\vec{x} - \vec{x}_s)$ is the Dirac delta function, $\vec{x}$ is an observation location, and $\vec{x}_s$ is a source location.

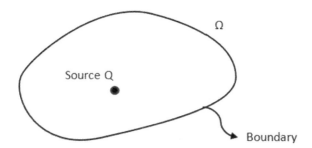

Fig. 9.1   Schematic diagram of computational acoustic domain.

Figure 9.1 shows the typical computational domain Ω with source strength Q.

If the acoustic domain of interest is outside the source domain, the homogenous form of the Helmholtz equation is appropriate to consider.

9.3   Kirchoff–Helmholtz Integral Equation

The Kirchoff–Helmholtz integral equation (KHIE) provides a framework for calculating the total field at a specific observation point based on the known properties of the incident field and the geometry of the scattering object.

The integral equation integrates over the boundary of the obstacle or aperture and involves Green's function associated with the Helmholtz equation [3]:

$$p(\vec{x}) = \frac{1}{4\pi} \iint_S \left\{ G(\vec{x}/\vec{x}_s) \frac{\partial p(\vec{x}_s)}{\partial n} - p(\vec{x}_s) \frac{\partial G(\vec{x}/\vec{x}_s)}{\partial n} \right\} dS \qquad (9.4)$$

$\vec{x}$, $\vec{x}_s$ are locations of observation and source points, $G(\vec{x}/\vec{x}_s)$ is a free space Green's function, which satisfies the inhomogeneous form of the Helmholtz equation. The general form of KHIE is,

$$p(\vec{x}) = \frac{1}{4\pi} \iint_S \{G(\vec{x}/\vec{x}_s)\nabla_{\vec{x}_s} p(\vec{x}_s) - p(\vec{x}_s)\nabla_{\vec{x}_s} G(\vec{x}/\vec{x}_s)\} \cdot \vec{n} dS(\vec{x}_s)$$
$$(9.5)$$

9.4 Acoustic Boundary Conditions

Boundary conditions in acoustics specify how sound waves interact with the boundaries of a domain. They play a crucial role in determining the behavior of sound waves within a given acoustic system. Depending on the nature of the problem and the physical setup, various types of boundary conditions can be applied.

Generally, three different types of boundary conditions are defined: Dirichlet, Neumann, and Robin boundary conditions [4]. They are mathematically expressed in arbitrary function $f(x, t)$ and bounded surface S as follows.

9.4.1 *Dirichlet condition (essential boundary conditions)*

$$f(x, t) = b(x, t) \quad \text{for } x \in S, \ t > 0 \tag{9.6}$$

The b function is prescribed in the Dirichlet boundary condition. If $b = 0$, then it is called a homogeneous Dirichlet boundary condition.

In acoustic analysis, the Dirichlet boundary conditions prescribe the sound pressure (or sometimes the particle velocity) at the domain's boundary. These conditions are often used when the sound field at the boundaries is under control.

Examples

• **Perfectly reflective boundary:** Sound waves are fully reflected at the boundary ($u = 0$).
• **Fixed pressure boundary:** The sound pressure is set to a specific value ($p = p_0$).
• **Fixed particle velocity boundary:** The particle velocity is set to a specific value ($u = u_0$).

9.4.2 *Neumann boundary conditions (natural boundary conditions)*

$$\nabla f(x, t) \cdot \vec{n} = b(x, t) \quad \text{for } x \in S, \ t > 0 \tag{9.7}$$

where $\vec{n}$ denotes the outer normal to boundary S. If the function $b = 0$ is called homogeneous boundary conditions. This condition represents that the normal gradient of the function is zero.

Neumann boundary conditions specify the normal derivative of the sound pressure (or particle velocity) at the boundary. These conditions represent conditions with a known acoustic energy at the boundary. The normal surface velocity is the preferred boundary compared to pressure in structural acoustic problems because vibration magnitude is insensitive to the background noise and easy to measure and compute from the numerical methods:

$$\vec{v}_n(\vec{x_\Omega}) = \vec{v} \cdot \vec{n} = \frac{-1}{jk\rho_o c} \nabla p(\vec{x_\Omega}) \tag{9.8}$$

where $\vec{x_\Omega}$ is on the boundary surface.

Examples

- **Sound-hard boundary:** The normal derivative of the sound pressure is zero, simulating a fully reflective boundary.
- **Sound-soft boundary:** The normal derivative of the sound pressure is proportional to the normal component of the particle velocity, simulating a perfectly transmitting boundary. The impedance of the surface material is very low compared to the surrounding acoustic medium for the soft boundary.

9.4.3 *Robin or mixed boundary conditions*

$$\alpha(x,t)f(x,t) + \beta(x,t)\nabla f(x,t) \cdot \vec{n} = b(x,t) \quad \text{for } x \in S,\ t > 0 \tag{9.9}$$

It is a mixed boundary condition. If β is zero, it represents Dirichlet conditions; if α is zero, it represents Neumann boundary conditions.

Impedance boundary conditions represent a relationship between the sound pressure and the normal particle velocity at the boundary. They are used to model surfaces with finite acoustic impedance.

Examples

- **Acoustic impedance boundary:** $p = Z_c u_n$, where Z_c is the characteristic impedance of the medium and u_n is the normal component of the particle velocity.
- **General impedance boundary:** $p = Z(\omega) u_n$, where $Z(\omega)$ is a frequency-dependent impedance.

9.4.3.1 *Mixed boundary conditions*

In some cases, a combination of different boundary conditions may be appropriate. For instance, you might have a surface with a combination of reflective and absorbing properties.

9.4.3.2 *Radiation boundary conditions*

Radiation boundary conditions are used to simulate sound waves propagating away from an unbounded domain. They are often applied at the exterior boundaries of a finite computational domain to mimic the behavior of sound waves radiating into the far field.

Radiation boundary conditions are particularly important when modeling problems where the sound field extends beyond the computational domain, such as in open-space or outdoor scenarios. They help prevent artificial reflections from the domain boundaries and ensure the simulated sound field behaves realistically.

There are also other types of boundary conditions commonly used in acoustic simulations. These are as follows.

9.4.3.3 *Perfectly matched layer boundary conditions*

Perfectly matched layer (PML) is a widely used technique to absorb outgoing waves effectively. It involves introducing an absorbing layer around the computational domain that gradually attenuates waves as they move away from the main domain. PML boundary conditions are especially useful for problems involving wave propagation in unbounded domains or when accurate absorption of waves is crucial.

9.4.3.4 *Sommerfeld radiation condition*

The Sommerfeld radiation condition is an asymptotic condition that describes how waves behave as they propagate away from a source. The acoustic field vanishes at the far away locations from the source. It involves specifying the decay rate of the sound pressure with distance from the source, typically proportional to $1/r$. The Sommerfeld boundary conditions are particularly useful for problems involving wave propagation in unbounded domains and are commonly used in both time-domain and frequency-domain simulations. In practice, implementing the Sommerfeld boundary conditions can be challenging due to the requirement of accurately approximating the behavior of waves in the far field.

The general form of the Sommerfeld radiation condition for an acoustic field (sound pressure) is [4]

$$\lim_{r\to\infty}\left[r\left(\frac{\partial p}{\partial r}+\frac{1}{c}\frac{\partial p}{\partial t}\right)\right]=0 \qquad (9.10)$$

where p is the sound pressure, r is the distance from the source point to the observation point, and c is the speed of sound.

9.4.3.5 *Symmetric boundary conditions*

Symmetric boundary conditions enforce symmetry or anti-symmetry of a physical quantity across a boundary. In other words, they specify that the value of the quantity at a boundary point is related to the value at its symmetric counterpart across the boundary.

Examples of symmetric boundary conditions in acoustics

- **Symmetric pressure boundary:** The normal derivative of sound pressure is zero along the boundary, preserving the symmetry of the pressure distribution.
- **Symmetric displacement boundary:** For a structure coupled with acoustics, symmetric displacement boundary conditions can be imposed to maintain symmetric vibrations.

- **Asymmetric boundary conditions:** Asymmetric boundary conditions allow sound waves to behave differently on either side of a boundary. When the problem involves asymmetry and want to capture the non-symmetric behavior accurately, then use these boundary conditions.

9.5 Types of Acoustic Problems and Their Classification

The acoustic problems can be broadly classified into two types based on the computational domain of interest. The first type of problem has a domain of interest external to the acoustic source, and the second one is interior problems.

9.5.1 *Exterior problem*

An exterior problem involves simulating a physical phenomenon in an unbounded domain or a domain extending infinitely in one or more directions. The simulation aims to capture the phenomenon's behavior as it propagates away from its source or interacts with distant boundaries.

Figure 9.2 shows the exterior domain with a surface S and normal pointing to the external and imaginary boundary at infinite, and Sommerfeld radiation boundary conditions represent it. The exterior problems use Dirichlet and Neumann boundary conditions.

Examples of exterior problems

Modeling sound radiation from a loudspeaker into open space, sound radiation from a vibrating structure with a known surface velocity, etc.

9.5.2 *Interior problem*

An interior problem involves simulating a physical phenomenon within a bounded domain. In this case, the region of interest is contained within a well-defined boundary, and the simulation focuses on how the phenomenon behaves within that boundary.

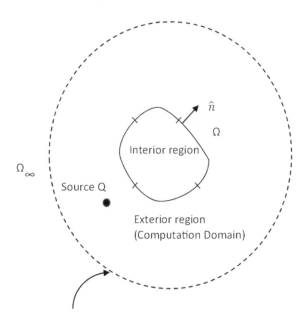

Imaginary boundary at infinite where Somerfield radiation applies.

Fig. 9.2   Exterior problem: computational acoustic domain with boundary conditions.

Figure 9.3 shows the interior domain volume (V_{int}) with a surface Ω; the interior region is the interested computational domain. The interior problems typically use the Dirichlet and Robin boundary conditions.

Examples of interior problems

Modeling sound propagation within a room with hard walls, acoustic modal analysis, etc.

Identifying whether a problem is an interior or exterior problem is crucial for selecting appropriate numerical methods, boundary conditions, and techniques. For example, exterior problems often require techniques like PMLs to effectively absorb outgoing waves, while interior problems might need absorbing boundary conditions or boundary layer meshing strategies. The choice of methods and approaches depends on the problem's classification as interior or exterior and the specific physics being modeled.

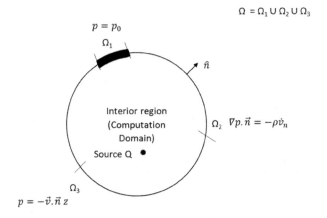

Fig. 9.3 Interior problem: computational acoustic domain with boundary conditions.

9.6 Approximate Solutions from Numerical Methods

The possible numerical solution of acoustic pressure for the governing equations (GEs) is expressed as

$$p(\vec{x}) = \sum_{i=1}^{N} a_n \phi_n \qquad (9.11)$$

a_n are unknown coefficients and ϕ_n are basis functions. The selection of the basis functions and calculation of the corresponding unknown coefficients leads to two methods for acoustic problems. The solution so obtained should satisfy the GE and boundary conditions (BCs). In the approximate solutions, the selection of basis function (Φ_n) either satisfies the GE or BCs and obtains the unweighted coefficients (a_n) by approximating the BCs or GE. FEM and BEM are the two most popular methods in acoustic problems, and the basis function selection for the same is described in Table 9.1.

Table 9.1. Selection of basis function in FEM and BEM methods.

	FEM	BEM
Basis function	Exactly satisfying the boundary conditions.	Satisfying the GE.
Unweighted coefficients	Are determined by approximating GE in the domain. Hence, the discretization is the domain.	Determined from approximating to BCs over the domain. Hence, the discretization is a boundary.
Discretization domain		
	3D problems: volume	3D problems: surface
	2D problems: surface	2D problems: lines, curves
	1D problems: lines, curves	

9.6.1 *Equations for undetermined coefficients*

Forming algebraic equations to calculate the undetermined coefficients from the GE and BCs is one of the important factors in the numerical solution accuracy. Various schemes have been developed over the years to determine the undetermined coefficients. A few of them are [4]:

- Collocation method,
- Subdomain method,
- Orthogonality method,
- Least square method,
- Rayleigh–Ritz method.

The simplest of these methods is the collocation method. The approximate solution should be satisfied at the specified locations or collocation points in the collocation method. In FEM, we satisfy the GE in the specified domain. Similarly, BEM satisfies the boundary conditions at specified locations. The collocation method suffers from numerical oscillation between the collocation points for complex problems.

The subdomain or partition method enforces that the solution should satisfy average values over the elements or subregions of the discretized domain. These methods required more computational time than the collocation method and were more effective in predicting global parameters than local quantities.

In the orthogonality method, linearly independent orthogonal functions are chosen so that the difference between the numerical solution and the specified boundary condition is orthogonal to each function. If the chosen orthogonal function is the same as the basis function, the numerical solution is called Galerkin's method. The computational time is similar to that in the subdomain method.

The least square method is based on minimizing the error between approximation and exact solution as a first step in calculating the error and minimizing the error with respect to undetermined coefficients. The convergence improves as the number of basis functions increases.

The Rayleigh–Ritz method is the least square method, but the minimization function is chosen according to the problem. The most preferable functions for minimization are the overall kinetic or potential energy.

9.7 Finite Element Method

The finite element method (FEM) is a powerful numerical technique used to solve various engineering and scientific problems, including acoustic and vibration analysis. It involves discretizing a complex domain into smaller, simpler elements with nodes at corners or along edges. Each element response is expressed in terms of a finite number of degrees of freedom, defined as the value of an unknown function at a set of nodal points. The solution to the GE is approximated from the collection of discrete models of all elements and satisfying the compatibility conditions between elements at the nodes.

9.7.1 *FEM for structural vibrations and acoustics*

Generally, two types of problems are considered. The first type of problem is obtaining the dynamics characteristics, such as natural frequencies and mode shapes, which requires solving eigenvalue problems. These types of problems are called free vibration analysis. The second type of problem is to predict the vibration response under periodic or transient loads, called a forced vibration problem. By assuming a time-harmonic response, the response solution for each frequency is obtained by solving the matrix equation directly. In the second approach, the response is expanded in terms of mode shapes obtained from the eigenvalue analysis.

For every element, the kinetic and potential energies are expressed in terms of the nodal displacement. The displacement throughout the element is obtained from simple interpolation functions regarding the nodal displacements or pressures. These simple interpolation functions are called shape functions, which are linear or quadratic. The stiffness and mass matrices of the elements are calculated for all the types of elements used in the discretization. The matrices for standard elements are readily available in commercial software. A simple explanation of the equation of motion for the discrete system in terms of the nodal degrees of freedom is given in the next section.

9.7.2 *Equation of motion for a discrete system*

The displacement across the element can be written in terms of the nodal degrees of freedom as [4]

$$\{u\} = [N]\{d\} \tag{9.12}$$

The vector of strain in terms of displacement can be written as

$$\{\varepsilon\} = [\Delta][N]\{d\} \tag{9.13}$$

The vector of stress in terms of strain can be written as

$$\{\sigma\} = [E][\Delta][N]\{d\} \tag{9.14}$$

Here, $[N]$ is shape function matrix, $\{d\}$ is nodal displacement vector, $[\Delta]$ is differential operator matrix and $[E]$ is a symmetric matrix of elastic constants, and the number of elements in the matrix is 36. Out of which, only 21 elements are independent because of symmetry. For isotropic materials, only two constants are independent: Young's modulus and Poisson's constant.

The single element kinetic and potential energies can be written in terms of nodal degree of freedom as

$$T = \frac{1}{2}\{\dot{d}\}^T \left(\iiint_V \rho [N]^T [N] dV(\vec{x}) \right) \{\dot{d}\} \qquad (9.15)$$

$$U = \frac{1}{2}\{d\}^T \left(\iiint_V [N]^T [\Delta]^T [E][\Delta][N] dV(\vec{x}) \right) \{d\} \qquad (9.16)$$

The kinetic and potential energy for all the elements (P) can be written as

$$T = \frac{1}{2}\{\dot{d}\}^T \left(\sum_{p=1}^{P} \iiint_V \rho [N]^T [N] dV(\vec{x}) \right) \{\dot{d}\} \qquad (9.17)$$

$$U = \frac{1}{2}\{d\}^T \left(\sum_{p=1}^{P} \iiint_V [N]^T [\Delta]^T [E][\Delta][N] dV(\vec{x}) \right) \{d\} \qquad (9.18)$$

The substitution of the total kinetic and potential energy terms into Lagrange's equation gives the discretized system equation of motion in terms of the stiffness and mass matrices, is shown as follows:

$$[M]\{\ddot{d}\} + [K]\{d\} = \{F\} \qquad (9.19)$$

where $[M]$ is the mass matrix, $[K]$ is Stiffness matrix, $\{F\}$ is force vector, $\{d\}$ is the nodal displacement vector.

The expressions for these matrices are

$$[M] = \sum_{p=1}^{P} \iiint_V \rho [N]^T [N] dV(\vec{x}) \qquad (9.20)$$

$$[K] = \sum_{p=1}^{P} \iiint_V [N]^T [\Delta]^T [E][\Delta][N] dV(\vec{x}) \qquad (9.21)$$

9.7.3 *Free vibration analysis*

The discrete equation of motion for modal analysis is obtained by removing the damping and external forces from Eq. (9.19), and it can be written as

$$[M]\{\ddot{d}\} + [K]\{d\} = \{0\} \qquad (9.22)$$

Assuming the time-harmonic and expressing the nodal displacement vector in terms of mode shape or eigenvector, $\{\emptyset\}$, as

$$\{d\} = \{\emptyset\} Re\{e^{j\omega t}\} \qquad (9.23)$$

Substituting Eq. (9.23) into Eq. (9.22) results in the free vibration analysis equation in the form of an eigenvalue problem, is shown as follows:

$$([K] - \omega^2[M])\{\emptyset\} = 0 \qquad (9.24)$$

The solution to Eq. (9.24) gives the eigen values and eigen vectors, which are natural frequencies and mode shapes, respectively. The number of eigenvalues extracted from the above equation is finite, but the actual dynamic systems are continuous and have infinite natural frequencies and mode shapes. However, most vibration analysis solvers require a few mode shapes to understand the dynamic behavior. Hence, the finite element analysis is an appropriate discrete numerical method for vibration analysis.

The mode shape represents the relative behavior, and the amplitude is arbitrary. So, normalization is applied to the mode shapes. The most popular ones are mass normalization and unit normalization. Mass normalization's advantage is obtaining the general mass and general stiffness, which can be useful for calculating the natural frequencies like the spring-mass system. Vibration modes are orthogonal, stiffness and mass matrices are symmetric, and generalized mass and stiffness matrices can be written as [4]

$$\{\emptyset_n\}^T[M]\{\emptyset_n\} = m_n$$
$$\{\emptyset_n\}^T[K]\{\emptyset_n\} = k_n \qquad (9.25)$$

The natural frequencies in terms of general mass and stiffness can be written as

$$\omega_n = 2\pi f_n = \sqrt{\frac{k_n}{m_n}} \tag{9.26}$$

9.7.4 *Forced vibration analysis*

The equation of motion with damping in terms of the nodal displacement vector can be written as

$$[M]\{\ddot{d}(t)\} + [C]\{\dot{d}(t)\} + [K]\{d(t)\} = \{F(t)\} \tag{9.27}$$

The input force excites the structure, and the equation of motion can be solved in the time as well as frequency domain. Based on the computational complexity in the time domain methods, the preferred method for solving the equation of motion is the frequency domain, assuming the time-harmonic. Assuming the displacement vector is time-harmonic because the input force is time-harmonic.

$$\{d(t)\} = \{d(\omega)e^{j\omega t}\} \tag{9.28}$$

Substituting Eq. (9.28) into Eq. (9.27) gives

$$(-\omega^2[M] + j\omega[C] + [K])\{d(\omega)\} = \{F(\omega)\} \tag{9.29}$$

The above equation is solved by combining all matrices on the left-hand side and solving the equation for the displacement vector at each frequency. This method makes the response results more accurate but time-consuming for many frequencies involving solution of the equation at each frequency.

The damping matrix $[C]$ can be obtained from the experimental data and expressed in terms of damping loss factor or proportional damping (Rayleigh damping).

9.7.4.1 *Modal summation method*

In this method, the nodal displacement vector is expressed in terms of mode shape function and time-harmonic as

$$\{d(t)\} = [\phi]\{\eta(\omega)\}Re\{e^{j\omega t}\} \tag{9.30}$$

where $[\phi]$ is a modal matrix with mode shape as a column vector, $\eta(\omega)$ is a modal participation factor as a function of frequency.

The expressions for general mass and stiffness are given in Eq. (9.25). The generalized damping matrix equation for assuming the proportional stiffness damping is written as

$$\{\emptyset_n\}^T[C]\{\emptyset_n\} = \zeta_n k_n \qquad (9.31)$$

where ζ_n is nth mode damping loss factor.

The equation of motion for the mode summation method to calculate the vibration response in terms of generalized mass, stiffness, damping, and mode participation factor can be written as

$$(-\omega^2 m_n + j\omega\zeta_n k_n + k_n)\eta_n = f_n \qquad (9.32)$$

where f_n is a generalized force for the nth mode. The mode participation factor is calculated from Eq. (9.32). The calculated mode participation factor is substituted into Eq. (9.30) to calculate the displacement vector. The selection of the number of modes in the expansion decides the calculation accuracy, and it is suggested to include the modes that exist twice the maximum interested frequency for the response analysis.

$$\eta_n = \frac{F_n}{(-\omega^2 m_n + j\omega\zeta_n k_n + k_n)} \qquad (9.33)$$

Pre-processing is one of the crucial steps while carrying out a numerical analysis. Therefore, an elucidated explanation is provided here. In the pre-processing stage, the solid model will be discretized into several elements, global matrices will be formed, material properties and boundary conditions will be applied, and an appropriate solver will be chosen to conduct the analysis. An overview of vibration analysis is shown in Fig. 9.4.

9.8 Boundary Element Method

The starting point for the BEM is the KHIE, and the general form of the boundary integral equation is given as [5]

$$\varepsilon(\vec{r_s})\emptyset(\vec{r}) = \oint_S \left[\emptyset(\vec{r})\frac{\partial G(\vec{r}/\vec{r_s})}{\partial n} - G(\vec{r}/\vec{r_s})\frac{\partial \emptyset(\vec{r})}{\partial n} \right] dS \qquad (9.34)$$

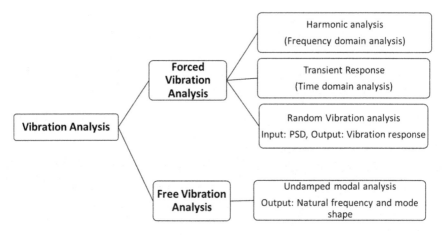

Fig. 9.4   Vibration analysis flow chart.

$$\varepsilon = \begin{cases} 1, & \overrightarrow{r_s} \in V_{\text{int}} \\ 0.5, & \overrightarrow{r_s} \in V_{\text{int}} \\ 0, & \overrightarrow{r_s} \in V_{\text{ext}} \end{cases} \qquad (9.35)$$

This method provides precise results for infinite or semi-infinite domains.

The BEM is advantageous because it only requires the calculations of boundary values instead of all the values in the domain defined by the partial differential equation. The acoustic pressure is estimated at the field points in the fluid by summing the contributions of simple acoustic sources (monopole and dipole) distributed over the elements on the boundary. The basis function satisfying the GE and numerical solution should find the multiplying undetermined coefficients of the basis function, which satisfies the boundary condition. BEM can be used for uncoupled as well as coupled problems. The terms $G(\vec{x}/\overrightarrow{x_s})$ and $\nabla_{\overrightarrow{x_s}}G(\vec{x}/\overrightarrow{x_s})$ in the KHIE are the solutions to GEs for unit amplitude monopole and dipole sources, respectively. The monopole and dipole sources are distributed over the boundary. The acoustic surface pressure and normal surface velocities are the exact solutions for the source coefficient in the BEM method.

The numerical form of the KHIE is written in the matrix form as [5]

$$([\boldsymbol{H}] - [\boldsymbol{I}])\{p\} = [\boldsymbol{G}]\left\{\frac{\partial p}{\partial n}\right\} \tag{9.36}$$

This equation is solved for pressure and normal pressure gradient at collocation points in the fluid for a given nodal pressure and normal velocities on the surface.

The BEM can be categorized into two main approaches: the direct boundary element method (DBEM) and the indirect boundary element method (IBEM). Both methods are used to solve the boundary value problems.

9.8.1 *Direct boundary element method*

The DBEM focuses on directly solving the integral equations derived from the GEs on the domain's boundary. It uses Green's function. The primary steps involved in the DBEM are as follows:

- Formulate the integral equations (e.g., Helmholtz Integral Equation or Burton–Miller Formulation) by converting the partial differential equation into an integral equation over the boundary.
- Discretize the boundary into small elements (panels or segments).
- Construct a linear system of equations by applying the discretized integral equations to the boundary elements.
- Solve the resulting system of linear equations to obtain the unknown boundary values, representing the solution to the problem.

The DBEM directly handles the integral equations on the boundary, making it conceptually straightforward. However, it requires solving a dense matrix system, which can become computationally demanding, especially for large-scale problems. DBEM is well-suited for problems with well-behaved integral kernels and where boundary discretization is relatively simple. Combined Helmholtz Integral Equation Formulation (CHIEF) and the Burton–Miller method are two different formulations to overcome the non-uniqueness problem. In the CHIEF method, collection points are included in the interior domain and convert the problem to an overdetermined system of equations [5]. Burton Miller's method uses the gradient Helmholtz integral formulation.

9.8.2 Indirect boundary element method

The KHIE is reformulated by using layer potentials in the IBEM. The potentials obtained on the boundary from the integral equation are used to calculate radiated acoustic pressure at the field points. The single layer potentials are continuous on the boundary, and its normal derivative jumps by the specified amount across the boundary. This method uses the Fredholm equations [5].

The comparison of the DBEM and the IBEM is presented in Table 9.2.

The BEM is a powerful numerical technique, but it comes with its own challenges like any computational method. Some of the common challenges associated with the BEM are as follows:

- **Singular integrals and near-field effects:** BEM often involves the evaluation of integrals over the boundary, and these integrals can become singular or highly oscillatory when evaluating points close to the boundary. Proper handling of singular integrals and near-field effects requires specialized techniques, such as singularity subtraction or regularization methods, to ensure accurate results. Due to singularity conditions, satisfying the boundary conditions on the surface at every point may not be feasible.

Table 9.2.   Comparison of the direct BEM and the indirect BEM.

DBEM	IBEM
Chosen for closed surfaces only	Chosen for open surfaces also
Non-symmetric matrix, not ideal for large size problem	Symmetric matrix and efficient for large mesh problems
Easy to interpret results. Point normal towards the acoustic domain	Interpretation results are complex. A jump of pressure conditions is required
The non-uniqueness problem for radiation problems	Nonexistence problem for closed problems

Hence, satisfying boundary conditions in an average sense may result in better solutions.

- **Ill-conditioning and matrix inversion:** When solving the linear equations resulting from the BEM, the system's matrix can become ill-conditioned, leading to numerical instability and accuracy issues. Efficient and robust matrix inversion and preconditioning methods are required to ensure reliable solutions.
- **The non-existence/non-uniqueness of basis functions** may occur at certain frequencies for surface monopole and dipole source distributions. At these frequencies, there is no radiation computationally. But it is not correct. There is a need to overcome this mathematical difficulty. Various methods are existing in the literature. However, a few methods to be mentioned are CHIEF, combination with the normal derivative formulation, etc.
- **Complex geometries:** Handling complex geometries with concave shapes, sharp corners, and intersecting boundaries can be challenging. The discontinuities in the normal velocity at the junctions lead to errors. Mesh generation and integration techniques must be carefully chosen to accurately represent the geometry and features.
- **Large-scale problems:** BEM can become computationally expensive for large-scale problems with many boundary elements. Efficient algorithms and parallelization techniques are essential to manage computational resources effectively.

Despite these challenges, the BEM remains a valuable tool for solving various engineering and scientific problems, offering advantages such as reduced dimensionality and the ability to model exterior domains efficiently. Researchers and practitioners continue to develop techniques and strategies to address these challenges and improve the accuracy and applicability of the BEM solutions.

9.9 Statistical Energy Analysis

SEA is a probabilistic energy-based method for analyzing complicated systems. The SEA emerged in the early 1960s to predict the vibration response of rocket noise [6]. Though the response of a structure can be predicted with a deterministic approach like FEM, the size of the models and computational speed was discouraging to the engineers. Deterministic methods are good for predicting lower-order modes. Traditionally, in the analysis of mechanical vibrations, the lowest modes are usually of most interest because these modes tend to produce a higher displacement response. However, while designing large and lightweight structures, it is imperative to account for high-frequency broadband loads to predict fatigue, failure, or noise emission. The computational time and size required for deterministic analysis at high frequencies are very discouraging. The number of modes per frequency bandwidth (modal density) increases at the high frequency. The SEA doesn't require the exact geometry of the system as it gives an ensemble average of the dynamic response of the system, and it can be focused on the energies of individual components but not on the spatial distribution of the responses. As SEA predicts the average response, it consumes a fraction of the time compared to the deterministic methods. Since its formulation, SEA has been widely used in many applications. It has also successfully predicted the average vibrational amplitudes and sound pressures in space vehicles, airplanes, ships, buildings, large machines, etc.

SEA starts by dividing the given system into subsystems. A subsystem is part of the system with "similar energy modes." The subsystem modes are assumed to have equal excitation with the forces. For example, a plate backed by a cavity is considered a structural-acoustic problem. The cavity is the subsystem with acoustic modes, and the structure is the subsystem with structural modes. Structural modes can still be divided into flexural, torsional, longitudinal, etc., which can be more subsystems. The primary assumption in SEA is that the modes in the subsystem are equally distributed with energy. Other assumptions in the SEA formulation are [6]:

- Subsystem response is not affected by the boundary conditions.
- Stored energy in the subsystem is associated with resonant modes. Energy is not created in the couplings between subsystems. The energy dissipation at the junction is included in the damping loss factor. The energy is equally partitioned among the modes.

- If there are no resonant modes in the frequency band, the SEA theory would not hold.
- The input power spectra are broadband and not in pure tone.

The concept of SEA is explained below with a simple two-subsystem example, as shown in Fig. 9.5. There are two subsystems. The arrow pointing towards the subsystems indicates the power received by the subsystems, while the arrows pointing away from the subsystems indicate the power lost from the subsystems. The arrows pointing within the subsystems indicate the exchange of power between the subsystems. W_1 and W_2 show the power entering the subsystems 1 and 2, respectively, whereas W_{1d} and W_{2d} indicate the power dissipated by the subsystems 1 and 2, respectively. W_{12} shows the power transfer from subsystems 1 to 2; similarly, W_{21} shows the power transfer from subsystems 2 to 1.

The following equations can show the dissipated power at an angular frequency ω in the subsystems [7]:

$$W_{1d} = \omega \eta_{1d} E_1 \qquad (9.37)$$

and

$$W_{2d} = \omega \eta_{2d} E_2 \qquad (9.38)$$

where η_{1d} and η_{2d} are damping loss factors of subsystems 1 and 2, respectively. E_1 and E_2 are the modes' total vibrational or acoustic energy at frequency f.

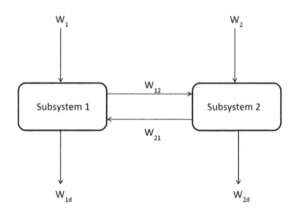

Fig. 9.5 SEA model of the interaction between two subsystems.

The net power transmitted between the subsystems can expressed as

$$W_{12} = \omega \eta_{12} E_1 \tag{9.39}$$

and

$$W_{21} = \omega \eta_{21} E_2 \tag{9.40}$$

where η_{12} and η_{21} are the coupling loss factors between the subsystems 1 and 2. The SEA calculation is based on the energy flow equilibrium; hence, the power balances for the two subsystems can be given by

$$W_1 + W_{21} = W_{12} + W_{1d} \tag{9.41}$$

and

$$W_2 + W_{12} = W_{21} + W_{2d} \tag{9.42}$$

When combining the above equations, the power balance equation for the two subsystems can be expressed in the matrix form as

$$\begin{bmatrix} W_1 \\ W_2 \end{bmatrix} = \omega \begin{bmatrix} (\eta_{1d} + \eta_{12}) & -\eta_{21} \\ -\eta_{12} & (\eta_{2d} + \eta_{21}) \end{bmatrix} \begin{bmatrix} E_1 \\ E_2 \end{bmatrix} \tag{9.43}$$

The generalized form of the power balance equation with n number of subsystems is

$$\begin{bmatrix} W_1 \\ W_2 \\ \cdots \\ W_n \end{bmatrix} = \omega \begin{bmatrix} \eta_1 & \eta_{21} & \cdots & \eta_{21} \\ -\eta_{12} & \eta_2 & \cdots & \cdots \\ \cdots & \cdots & \cdots & \cdots \\ -\eta_{1n} & \cdots & \cdots & \eta_n \end{bmatrix} \begin{bmatrix} E_1 \\ E_2 \\ \cdots \\ E_n \end{bmatrix} \tag{9.44}$$

where η_i stands for the total loss factor of the ith system, which is the summation of the damping loss factor and the coupling loss factors and, in general, can be stated as [7]:

$$\eta_i = \eta_{id} + \sum_{j=1, j \neq i}^{n} \eta_{ij} \tag{9.45}$$

where n is the number of subsystems, and the subscripts i and j represent the identities of the subsystems. Equation (9.44) is solved for subsystem energies for given input powers.

$$W = \omega \eta E \qquad (9.46)$$

$$E = \frac{\eta^{-1} W}{\omega} \qquad (9.47)$$

Typical SEA models consist of 20–200 subsystems. The computation time needed for obtaining the results is modest compared to deterministic calculations for the same frequency range. Based on the results for the energies of the subsystems, other variables, such as acceleration and sound pressure, can also be determined. Again, it must be kept in mind that these results are averaged over frequency and space.

With the energy of subsystems, one can calculate the acoustic subsystem's pressure and the structural subsystem's velocity [6].

For an acoustic cavity,

$$E = \frac{|p^2| V}{\rho c^2} \qquad (9.48)$$

where p is the acoustic pressure, V is the volume of the cavity, ρ is the density of the medium and c is the speed of sound in that medium.

For a structural subsystem,

$$E = \frac{1}{2} m v^2 \qquad (9.49)$$

where, m is the mass and v is vibration velocity.

9.9.1 *Procedure of SEA*

Figure 9.6 shows the general procedure of SEA that is followed to do the computations with the commercial software.

In summary, the required parameters to perform SEA are as follows:

1. Modal density,
2. Damping loss factor,

3. Coupling loss factor,
4. Power input.

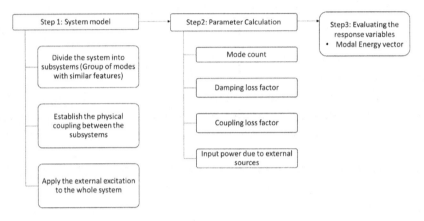

Fig. 9.6   Overview of the SEA.

9.9.2 *Modal density*

It is calculated using the wave approach, i.e., by knowing the phase and group velocity of the mechanical wave. The group velocity (c_g) is defined as the velocity of the wave with which the overall shape of the wave's amplitudes propagates through a medium. The phase velocity (c_p) is the rate at which the phase of the wave propagates. The various mode types for structure are bending, torsional, longitudinal, and transverse modes; similarly, for acoustic analysis, 1D, 2D and 3D mode groups are used for acoustic space. The modal density is higher for larger structures than smaller ones, and accuracy in the SEA model at lower frequencies is better for large structures. The mathematical expressions of the group velocity (c_g) and phase velocity(c_p) is given by [6]

$$c_g = \frac{\partial \omega}{\partial k} \tag{9.50}$$

$$c_p = \frac{\omega}{k} \tag{9.51}$$

Where k is the wave number. The expressions for modal density in terms of 1D, 2D, and 3D are tabulated as follows:

Dimensionality	Modal density expression
1D	$n(\omega) = \dfrac{L}{\pi c_g}$, L is length, c_g is group speed
2D	$n(\omega) = \dfrac{A\omega}{2\pi c_p c_g}$, A is Area, c_p is phase speed
3D	$n(\omega) = \dfrac{V\omega^2}{2\pi^2 c^3} + \dfrac{A\omega}{8\pi^2 c^2} + \dfrac{P}{16\pi c}$ P is the perimeter, c is the speed of sound

9.9.3 *Damping loss factor*

Energy dissipation at the junctions and within the subsystem is expressed as a damping loss factor. The damping loss factor is the same for all subsystem modes in the chosen frequency band. Mostly, the damping loss factor is known from measurements or empirical equations. It is measured from the decay rate, power injection, and experimental SEA in Table 9.3.

Decay rate method: This method provides impulse to the subsystem and measures the time domain response. The damping loss factor is calculated from the measured reverberation time (T_{60}) as follows:

$$\eta = \frac{2.2}{fT_{60}} \tag{9.52}$$

Table 9.3. Damping loss factor calculation for various input parameters.

Input parameter	Damping loss factor
Acoustic absorption coefficient ($\bar{\alpha}$)	$\eta = \dfrac{Ac}{4\omega V}\bar{\alpha}$, V is the cavity volume A is the total cavity surface area
Reverberation time (T_{60})	$\eta = \dfrac{2.2}{fT_{60}}$, f is frequency
Critical damping ratio (ζ)	$\eta = 2\zeta$
Band-averaged damping loss factor	$\eta = \dfrac{1}{N}\sum_{\omega_r \in \Delta\omega} \eta_r$

Power injection method: This method is based on the measured input power P and response energy E, and calculates the damping as follows:

$$\eta = P/\omega E \tag{9.53}$$

9.9.4 *Coupling loss factor*

The coupling loss factor is an essential parameter in SEA. It is associated with the energy transmitted from one subsystem to another. The formulation of coupling loss factors depends on the type of junctions and the properties of the subsystems. There are two approaches for calculating the junction coupling loss factor: the modal approach and wave approach. The wave approach considers the power transmission for semi-infinite structures, and it is easy to implement compared to the modal approach with finite structures.

The coupling loss factors in both directions are related to modal density and establishes a consistent relationship.

$$\eta_{ij} n_i = \eta_{ji} n_j \tag{9.54}$$

where n_i, n_j represents the modal density. The modal density presents the number of modes per unit frequency. It is essential to calculate the unknown coupling loss factor of a particular subsystem by means of the consistency relationship. If the coupling loss factors in both directions of the subsystems in the model are known, then evaluating the modal density may not be a requisite.

The generalized equation for the coupling loss factor, which applies to all the junctions is [7]

$$\eta_{12} = \frac{\tau_c}{2\pi f n_1} = \frac{\tau_c}{\omega n_1} \tag{9.55}$$

where τ_c is the transmission coefficient between subsystem 1 and subsystem 2, and n_1 is the modal density of subsystem 1. The transmission coefficient (τ_c) depends on the type of subsystems connected.

The SEA is focused on calculating the responses for resonant transmission between subsystems. The SEA model's basic assumption is that each subsystem's energy is contained in the resonant modes so that the energy is proportional to the damping. However, the response of an element is sometimes not proportional to the damping. In this case, the excited behavior is non-resonant. For coupling between two subsystems separated by a plate or a wall, and there is a transmission through the plate at a frequency below the resonant frequency of the plate, the transmission can be termed as non-resonant transmission.

9.9.5 *Power input*

The power input to the subsystems in the SEA models is given as user input. The SEA model may calculate the input power from the applied forces, moments, velocities, or acoustic pressures. Power measurements can also be applied as input in the model.

9.10 Structural-Acoustic Coupling

The structural-acoustic coupled problem has been of great interest to many researchers for the several decades. In the 1960s, Warburton was one of the first researchers who found that a cylindrical shell containing air as an acoustic fluid possesses natural frequencies close to either uncoupled structural natural frequencies or the uncoupled acoustic natural frequencies of the enclosed air. Dowell *et al.* [8] expanded upon this idea by developing a theoretical model that combines the uncoupled acoustic cavity modes and the uncoupled structural modes into a system of coupled equation. This model serves as the theoretical framework to investigate structural-acoustic interaction in various systems of practical interest, like rectangular enclosures and resonant modes of guitar bodies.

Figure 9.7 shows an irregular acoustic cavity enclosed in a structural interface. The flexible structural surface is S_f.

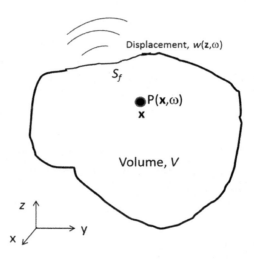

Fig. 9.7   Schematic diagram of structural-acoustic coupling system.

The general equation of inside cavity pressure P at some location and the compliance wall vibration velocity w at some location on flexible surface for the uncoupled cavity modes N and structural modes M are [9]:

$$P(\vec{x}, \omega) = \sum_{n=1}^{N} a_n(\omega)\psi_n(\vec{x}) = \boldsymbol{\Psi}^{\mathbf{T}}\mathbf{a} \qquad (9.56)$$

$$w(\vec{z}, \omega) = \sum_{m=1}^{M} b_m(\omega)\phi_m(\vec{z}) = \boldsymbol{\phi}^{\mathbf{T}}\mathbf{b} \qquad (9.57)$$

ψ_n is the uncoupled acoustic mode shape function, a_n is the complex amplitude of the nth acoustic pressure mode, Φ_m is the uncoupled vibration mode shape function, and b_m is the complex amplitude of the mth vibration velocity.

Here, normal modes satisfy the properties of orthogonality for uncoupled acoustic modes [9],

$$\frac{1}{V}\sum \psi_i\psi_j = 0 \quad \text{for } i \neq j$$

$$= M_n \quad \text{for } i = j \qquad (9.58)$$

Similarly, for uncoupled structural modes,

$$\sum \phi_l \phi_k = 0 \quad \text{for } l \neq k$$
$$= M_m \quad \text{for } l = k \qquad (9.59)$$

The GE for the coupled structural-acoustical problem is given by

$$[M] \begin{Bmatrix} \ddot{a}_n \\ \ddot{q}_k \end{Bmatrix} + [G] \begin{Bmatrix} \dot{a}_n \\ \dot{q}_k \end{Bmatrix} + [K] \begin{Bmatrix} a_n \\ q_k \end{Bmatrix} = \begin{Bmatrix} 0 \\ 0 \end{Bmatrix} \qquad (9.60)$$

$$[M] = \begin{bmatrix} VM_n & 0 \\ 0 & M_m \end{bmatrix}, \quad [G] = \begin{bmatrix} 0 & -S_f c^2 [C_{m,n}] \\ S_f c^2 [C_{m,n}]^T & 0 \end{bmatrix},$$

$$[K] = \begin{bmatrix} VM_n \Omega_n^2 & 0 \\ 0 & M_m \Omega_m^2 \end{bmatrix}.$$

where $[M]$, $[K]$ and $[G]$ are mass matrix, stiffness matrix, and coupling matrix, respectively. Here, Ω_n and Ω_m are uncoupled acoustic and structural natural frequencies. S_f is flexible surface area; c is the velocity of sound in the fluid, and ρ_0 is the structural mass density. $C_{m,n}$ is the uncoupled structural-acoustic mode shape coefficient. It is the relationship between the uncoupled structural and acoustic mode shape of vibration surface S_f and its measure of spatial match between the structure and acoustic mode. When the value of the coupling coefficient is unity, then the two modes are in the exact spatial match, and when the value is zero, there is no spatial match between the two modes.

Hence, $C_{m,n}$ is given as [9]

$$C_{m,n} = \frac{1}{S_f} \sum \psi_n \phi_m \text{ over the area } S_f.$$

A transfer factor is used to identify the well-coupled modes. The transfer factor for thin cavity flexible wall for the nth cavity mode and mth structural mode can be written as [9]

$$T_{n,m} = \left(1 + \frac{(\omega_n^2 - \omega_m^2)\rho_s h S_f V}{4\rho_0 c_0^2 C_{n,m}^2} \right)^{-1} \tag{9.61}$$

If $|T_{n,m}| \cong 1$, then the cavity mode and structural mode are well coupled, and the energy involved in the interaction is comparable with that of the constituent subsystems. However, if $|T_{n,m}| \ll 1$ then, the modal interaction is small, and the system is said to be weakly coupled.

An example of a structural acoustic problem is computation of the breakout noise from a rectangular plenum [10]. Figure 9.8 shows a flow chart containing the numerical methodology adopted for the breakout noise analysis.

In the FEM part, uncoupled structural and acoustic modal analysis is carried out to determine the uncoupled natural frequencies and mode shapes of the plenum wall and acoustic cavity. After that, the coupled modal analysis of the system is carried out to find the coupled modal parameters. By providing an input excitation (noise source), the response of the coupled system is found using the FEM. The FEM results are then given as input in the BEM analysis to carry out the breakout sound radiation analysis.

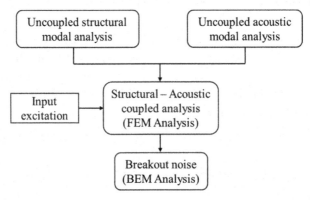

Fig. 9.8   Flow chart showing the numerical methodology for breakout noise analysis.

9.11 Numerical Modeling

The structural-acoustic problems are solved by discrete methods such as the FEM, BEM, and the SEA. However, discrete methods are precise only at lower frequencies where resonant mode response is predominant and has a lower modal density, which occurs at lower frequencies. The theoretical aspects were discussed in the previous sections. A good number of commercial software programs are available for discrete methods, such as ANSYS, ACTRAN, Simcenter 3D. The VA One software is known for the SEA methods, and this software incorporates both the discrete and SEA methods to cover the low and high frequencies [11–15].

Figure 9.9 shows the typical vibro-acoustic response of a structure as a function of frequency. It can be observed that at low frequencies, the modal density is low, and modes are well separated. The modal density is higher at higher frequencies, and the energy-based algorithms are appropriate. Hybrid methods need to be used for mid-frequencies [15].

The current trend in commercial software is to integrate the various modules and solve multi-physics problems on a single platform to minimize user interaction time. This section gives an overview

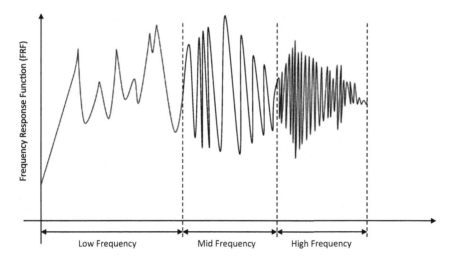

Fig. 9.9   Typical vibro-acoustic response as a function of frequency.

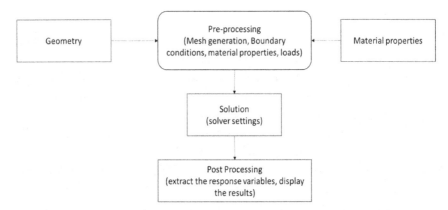

Fig. 9.10 The process map for acoustic radiation prediction in commercial software.

of the numerical model implementation in the software. Significant changes are taking place with the user interface and integration of the modules by adding artificial intelligence and machine learning algorithms. This section provides an overview of numerical modeling with commercial tools but not the usage of tools. Hence, the user manual of commercial software is a good start to the implementation. The major steps in the discrete methods are (i) pre-processing, (ii) solver and (iii) post-processing. The pre-processing and post-processing time requirements may vary according to the problem and details to be retained for the solver. The major functions involved in pre-processing are creating the proper geometry, meshing with the appropriate element type, assigning the material properties to the meshed model, and applying the loads. The solver step involves choosing the appropriate parameters for solver settings and solving the problem according to the domain of interest and frequency range. The last step is to extract the results in the required form as tables, charts, and contour maps. The typical steps used in commercial software for structure-borne noise and airborne noise are given in Fig. 9.10.

9.11.1 *Pre-processing*

The primary objective in pre-processing is to discretize the geometry into a mesh grid pattern with the proper elements (connecting

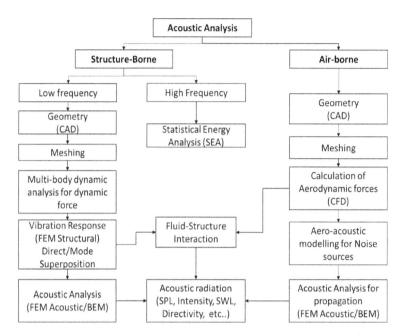

Fig. 9.11   Finite element analysis process flowchart in software.

various nodes), which is called meshing. The geometric and material properties are assigned to the mesh. Then, boundary conditions and external excitation are applied to the meshed model for solving.

Figure 9.11 shows the steps involved in the FEM analyses using the software as a flow chart.

9.11.2 *Meshing*

Creating a suitable mesh for acoustic simulations is essential for accurate and reliable results. The mesh quality significantly influences numerical simulations' accuracy, convergence, and computational efficiency.

9.11.2.1 *Element types*

Choose appropriate element types based on the geometry and physics of the problem. The various elements considered for acoustic and vibration analysis are line, surface, and solid elements, as shown in Table 9.4. For acoustics, commonly used elements include triangles

Table 9.4.   Various elements used in vibration and acoustic analysis.

S.No.	Type	Description
1	**Line element** One dimension is large as compared to other dimensions $(x \gg y, z)$ Node 1 Node 2	• It has two nodes for linear elements and three nodes for quadratic elements. • Beam line element (Bending, compression and torsion can be included). • Rod line element (Bending is not included). • Spring element can be modeled as line element (not included mass effects).
	Element type Various line elements used in the analysis are Rod, bar, spring, pipe, beam.	**Inputs** • Length, cross-section is specified. • Material Properties (density, shear modulus, damping loss factor).

2	**Surface element**
	 Two dimensions are large as compared to other dimensions $(x, y \gg z)$
	
	 Tri
	 Linear Elements Quadratic Elements
	
	 Quad
	 Linear Elements Quadratic Elements
		• Number of nodes for linear surface elements are:
		    o Triangular→ 3
		    o Quad→ 4
		 • Number of nodes for Quadratic surface elements are:
		    o Triangular→ 6
		    o Quad→ 8
		 • The type of stresses can be included in the plate elements are.
		 • Membrane in-plane stress.
		 • Bending and Twisting moments.
		 • Transverse shear.
		 • Shell elements are plate elements but also support coupling between bending and membrane stress due to curvature.

Table 9.4. (*Continued*)

S.No.	Type	Description
	Element type Various surface elements used in the analysis are Triangular, Quad etc.,	**Inputs** • Thickness. • Material Properties (density, shear modulus, damping loss factor).
3	**Solid elements** All the dimensions are in the same order $(x{\sim}y{\sim}z)$.	• Normal and shear stresses in three directions are calculated. • Transition nodes can be added to the linear elements.

| Linear Elements | Quadratic Elements |

Element type Solid elements are hexahedral (brick), wedge, tetrahedral, pyramid, etc.

(2D), tetrahedra (3D), and possibly quadrilaterals (2D) or hexahedra (3D) for more complex problems, as shown in Table 9.4.

9.11.2.2 *Element size and density*

The minimum number of elements per wavelength in acoustic analysis for linear shape function is six, as shown in Fig. 9.12. At higher frequencies, the element size decreases because of the smaller wavelength. Hence, the computational requirements increase for higher frequencies and large-size structures. In general, the element size is

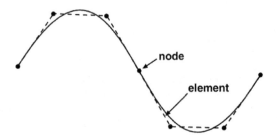

Fig. 9.12 Number of linear elements to fit the sine wave.

chosen to satisfy the following in equality:

$$e_{max} \leq \frac{\lambda}{6} \tag{9.62}$$

where λ is the wavelength of the largest frequency of interest in the simulation.

Smaller elements near sources, boundaries, and regions of interest can capture local details accurately. Gradually increase element size away from the sources to represent the far field. Use adaptive mesh refinement to concentrate elements with high gradients, ensuring an accurate representation of the wave behavior.

9.11.2.3 *Mesh quality metrics*

Use mesh quality metrics like aspect ratio, skewness, and Jacobian determinants to assess the quality of elements. Avoid highly distorted or degenerate elements. Aim for higher-quality elements (closer to equilateral triangles or regular tetrahedra) to improve simulation accuracy and convergence.

9.11.2.4 *Size gradation*

Use gradual size gradation to transition from fine to coarse elements, ensuring a smooth mesh transition and accurate representation of wave behavior.

9.11.2.5 *Domain symmetry*

Exploit symmetry or periodicity in the problem to reduce computational costs and simplify meshing.

9.11.2.6 *Number of elements and degrees of freedom*

A proper balance between mesh refinement and computational resources needs to be maintained. Too fine a mesh might lead to excessive computation time. Monitor the number of elements and degrees of freedom as you refine the mesh to ensure feasible simulation times.

9.11.2.7 *Sensitivity analysis*

Perform sensitivity analysis by varying mesh parameters (element size, density, etc.) to assess how changes affect simulation results.

The typical information to be considered while selecting the elements for acoustic analysis are number of nodes, degree of freedom at each node, element shape, fluid-structure interaction capability, symmetric/unsymmetric, perfect matched layer (PML) capability, and non-uniform acoustic media properties.

9.11.3 *Boundary conditions*

For vibration problems, the boundary conditions are defined in terms of displacement and rotation at nodes.

Type of boundary condition	Remarks
Clamped	All DOFs are constrained.
Hinged	Only displacements are constrained.
Free	No DOFs are constrained.

Typical boundary conditions for acoustic analysis are as follows:

Type of boundary condition	Remarks
Rigid	Velocity is zero
Vacuum	Pressure is zero
Absorption	Impedance
Anechoic termination	Characteristic impedance
Perforated sheet	Transfer admittance
Waveguide open to the atmosphere (for example, duct termination to atmosphere)	Radiation impedance

Most commercial software considered acoustic rigid boundary conditions as default conditions on the outer surface in the FEM acoustic simulation.

External excitations: A few examples of external excitation for vibration analysis are given in Table 9.5, and excitations for acoustic analysis are given in Table 9.6.

9.12 Introduction to Aeroacoustics

The science of flow-induced sound is called aeroacoustics, and the source is fluid flow. The fluid-borne sound has tonal and broadband characteristics. This sound is dominant at higher frequencies. Vortex shedding frequency, spatially non-uniform flow, is one of the most prominent sources of tonal noise. Turbulence in fluid, flow separation,

Table 9.5. External excitations for vibration analysis.

Type of load	Loading pattern
Constant	
Periodic	
Random	
Impulse	

Table 9.6.   External excitations for acoustic analysis.

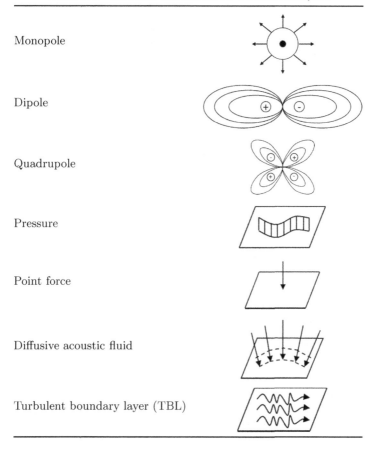

Monopole	
Dipole	
Quadrupole	
Pressure	
Point force	
Diffusive acoustic fluid	
Turbulent boundary layer (TBL)	

unsteady inflow, and forces on surfaces are responsible for broadband noise.

Aerodynamic noise results from unsteady gas flow and the interaction of the unsteady gas flow with the associated structure. The unwanted gas flow and structure interaction may cause serious problems in industrial products, such as the instability of the structures and structure fatigue. Accordingly, simulating the aerodynamic noise is necessary at the design stage. However, due to the nature of turbulent flow and the computational power limitation, obtaining a reliable unsteady (transient) CFD solution for the aerodynamic noise analysis is not always feasible. The computational effort and time are major hindrances. Even if there were no time limitations, any of the commonly used turbulent models cannot solve all turbulence scales [16]. Therefore, a time-efficient method with acceptable accuracy is needed to estimate flow noise. Several well-known theories, such as the theory of Lighthill [17, 18], Curle's theory [16], and the theory of Ffowcs Williams and Hawkings (FWH) [19] have been successfully applied to aeroacoustics problems. The theory of Lighthill is the foundation of the FWH approach. The fluid characteristic dimensions, length scale, and acoustic wavelength are at different levels. Similarly, the fluid energy and acoustic energy are in different order. Hence, solving the combined fluid dynamics and acoustic equations is quite a challenge in computational aeroacoustics.

9.12.1 *Acoustic analogy*

Analogy gives the reformulation of known equations. The fluid dynamic equations are reformulated as inhomogeneous wave equations in the aeroacoustics analogies. A brief description of the Lighthill analogy, Curle's model, and FWH formulation are given in this section.

9.12.1.1 *Lighthill's analogy*

Michael James Lighthill published a paper in 1952 for predicting sound generated by aerodynamic flow sources from the momentum equation as an inhomogeneous wave equation [17]. His research work is the birth of Aeroacoustics. After that, aeroacoustics has become a branch of acoustics that studies the sound induced by aerodynamic activities or fluid flow. In 70 years, the theory of aeroacoustics has been greatly developed and widely applied in modern engineering. The subject of Lighthill's paper is sound generated aerodynamically, a byproduct of airflow distinct from the sound produced by the vibration of solids.

His formulation has two major assumptions: (i) Acoustic propagation of fluctuations in the flow is not considered, (ii) Back-reaction of the sound produced on the flow field itself is precluded.

Therefore, the effects of solid boundaries are neglected. However, the back-reaction is only anticipated when a resonator (i.e., a cavity) is close to the flow field. Accordingly, his theory applies to most engineering problems. Furthermore, his theory is confined to subsonic flows and should not be used to analyze the transition to supersonic flow. Lighthill examined a limited volume of a fluctuating fluid flow in a very large fluid volume. The remainder of the fluid is assumed to be at rest. He then compared the equations governing the density fluctuations in the real fluid with a uniform acoustic medium at rest, which coincides with the real fluid outside the flow region. A force field is acquired by calculating the difference between the fluctuating and stationary parts. This force field is applied to the acoustic medium, and then acoustic metrics can be predicted away from the source by solving the Helmholtz equation. Helmholtz equation can be solved easily if a free field is assumed or solved using numerical simulation. This analogy has two significant advantages, as mentioned in his paper. First, since we are not concerned with the back-reaction of the sound on the flow, it is appropriate to consider the sound as produced by the fluctuating flow after the manner of a forced oscillation. Secondly, it is best to take the free system, on which the forcing is considered to occur, as a uniform acoustic medium at rest. Otherwise, it would be necessary to consider the modifications due to convection with the turbulent flow and wave propagation at different speeds, which would be difficult to handle. Using the method just described, an equivalent external force field is used to describe the acoustic source generation in the fluid [17].

$$\frac{1}{c_0^2}\frac{\partial^2 p'}{\partial t^2} - \frac{\partial^2 p'}{\partial x_i^2} = \frac{\partial^2}{\partial x_i \partial x_j}(\rho v_i v_j - \sigma_{ij}) - \frac{\partial f_i}{\partial x_i} + \frac{\partial^2}{\partial t^2}\left(\frac{p'}{c_0^2} - \rho'\right)$$

(9.63)

The perturbations of pressure and density defined by $p' = p - p_0$ and $\rho' = \rho - \rho_0$.

The pressure and density fluctuations are applicable in linear acoustic space around the listener and in the nonlinear source region.

The first term on the right-hand side of the equation represents the effect of viscous stress induced by molecular transport of momentum (σ_{ij}), and Reynold stress ($\rho v_i v_j$) includes the nonlinear convection of momentum. The second term represents the external force. The third term on the right-hand side represents the generalization of the entropy production, and it includes the complex effects due to the convection of entropy non-uniformities.

Lighthill made critical assumptions that if the external force and entropy term contribution are negligible, then the source of sound is flow only. For higher velocities, the viscous effects are negligible. Hence, the sound source is nonlinear convective effects. Equation (9.63) can be solved by the integral formulation using Green's function.

Lighthill has demonstrated that the sound produced by a free turbulent isentropic flow has the character of a quadrupole and showed that the radiated power is proportional to the eighth power of flow velocity. A better way of putting it is that since, in such flows, there is no net volume injection due to entropy production nor any external force field, the sound field can, at most, be quadrupole.

9.12.1.2 *Ffowcs Williams-Hawking formulation*

The curle formulation is an extension of Lighthill's formulation for flows in the presence of wall section S. In contrast, the formulation of FWH allows the use of a moving control surface $S(t)$. FWH formulation has vast applications in rotor dynamics applications, such as propeller noise, helicopter blade noise, and fan noise. This equation is for perturbed density with three source terms: Quadrupole source in terms of Lighthill stress tensor, loading, and thickness terms. Thickness noise is created due to the motion of the surface in the normal direction. Loading noise is associated with pressure distribution on the surface.

Once the turbulent field is fully modeled, the acoustic sources are computed on the surface of the fluid domain using FWH methodology, which is the most general form of Lighthill's analogy. This method computes the acoustic sources from the flow field in terms of monopoles, dipoles, and quadrupoles. The FWH equation is given as [19]

$$\frac{1}{c^2}\frac{\partial^2 p'}{\partial t^2} - \frac{\partial^2 p'}{\partial x_i x_j} = \frac{\partial^2}{\partial x_i x_j}\{T_{ij}H(S)\}$$

$$-\frac{\partial}{\partial x_i}\{[P_{ij}n_j + \rho u_i(u_n - v_n)]\delta(S)\}$$

$$+\frac{\partial}{\partial t}\{[\rho_0 v_n + \rho(u_n - v_n)] + \delta(S)\} \quad (9.64)$$

where u_i is the flow velocity component in the x_i direction, u_n is the flow velocity component normal to the surface $S = 0$, v_i is the surface velocity component in the x_i direction, v_n is the surface velocity component normal to the surface, $\delta(S)$ is the Dirac delta function and $H(S)$ Heaviside function.

The Lighthill stress tensor T_{ij} is defined as

$$T_{ij} = \rho u_i u_j + P_{ij} - c^2(\rho - \rho_0)\delta_{ij} \quad (9.65)$$

The solution to the FWH equation (9.64) is obtained using the free-space Green's function $(\delta(G)/4\pi r)$. The complete solution consists of surface integrals and volume integrals. The surface integrals represent the contributions from monopole and dipole acoustic sources and partially from quadrupole sources. In contrast, the volume integrals represent quadrupole (volume) sources in the region outside the source surface. The contribution of the volume integrals becomes negligible when the flow is low subsonic, and the source surface encloses the source region.

9.12.1.3 *Implementation*

Commercial codes such as STAR CCM+ [20], Fluent [21] and LMS virtual lab [14] have incorporated the FWH approach in a computational aero-acoustics module, which assumes that obstacles exist between the sound sources and the receivers [19]. Therefore, the sound radiation problem is inherently a weak part of the simulation, especially if the sound source is in a waveguide or duct, enclosed, or obstructed in some way. The various unsteady CFD methods used for estimating the aerodynamic forces are [20, 21] as follows:

- **Direct numerical simulations (DNS):** Flow and acoustic fields are solved from the basic differential equations. It is computationally very expensive.
- **Large Eddy simulations (LES):** Large-scale eddies are resolved, and basic equations are filtered. Still, these methods are computationally expensive.
- **Unsteady Reynolds averaged Navier–Stokes simulation:** The turbulence is completely modeled and calculated in the ensemble averaging of the basic equations. The computational cost is independent of the Reynolds number.

References

[1] Munjal, M. L., *Acoustics of Ducts and Mufflers*, Second Edition, Wiley, Chichester, UK (2014).

[2] Kreyszig, E, *Advanced Engineering Mathematics*, Eighth Edition, Wiley (2006).

[3] Norton, M. P., *Fundamentals of Noise and Vibration Analysis for Engineers*, Cambridge Press (1996).

[4] Noureddine, A. and Franck, S., *Finite Element and Boundary Methods in Structural Acoustics and Vibration*, First Edition, CRC Press (2017).

[5] Stephen, K., The boundary element method in acoustics: A survey, *Applied Sciences*, 9, p. 1642 (2019).

[6] Lyon, R. H. and Dejong R. G., *Theory and application of Statical Energy Analysis*, Second Edition, Boston, MA, Buttersworth-Heimemann (1995).

[7] Yoganandh, M., Jade, N. and Venkatesham, B., Prediction of insertion loss of lagging in the rectangular duct using statistical energy analysis, *Noise Control Engineering Journal*, 67(6), pp. 438–446 (2019).

[8] Dowell E. H., Gorman G. F. and Smith, D. A., Acoustoelasticity: General theory, acoustic modes and forced response to sinusoidal excitation, including comparisons with experiment, *Journal of Sound and Vibration*, 52, pp. 519–542 (1977).

[9] Venkatesham, B., Breakout noise from the coupled acoustic-structural HVAC systems, PhD Thesis, Indian Institute of Science, Bangalore (2008).

[10] Venkatesham, B., Tiwari, M. and Munjal, M. L., Analytical prediction of the breakout noise from a reactive rectangular plenum with four flexible walls, *The Journal of the Acoustical Society of America*, 128(4), 1789–1799 (2010).

[11] ANSYS User Manual, ANSYS Inc. (2016).

[12] ACTRAN User Manual, Free Field Technologies, Hexagon AB (2020).

[13] Acoustic User's Guide, Siemens Product Lifecycle Management Software Inc. (2019).

[14] User Manual, LMS Virtual Lab Acoustic, LMS International (2015).

[15] VA One User manual, ESI (2015).

[16] Manfred, K., *Theoretical Acoustics Part: Aeroacoustics*, Lecture Notes (2021).

[17] Lighthill, M. J., On sound generated aerodynamically I. General theory, *Proceedings of the Royal Society of London. Series A. Mathematical and Physical Sciences*, 211, 564–587 (1952).

[18] Lighthill, M. J., On sound generated aerodynamically. II. Turbulence as a source of sound, *Proceedings of the Royal Society of London. Series A. Mathematical and Physical Sciences*, 222, 1–32 (1954).

[19] Ffowcs Williams, J. E. and Hawkings, D. L. Sound generation by turbulence and surfaces in arbitrary motion, *Philosophical Transactions of the Royal Society of London. Series A, Mathematical and Physical Sciences*, 264(1151), 321–342 (1969).

[20] STAR CCM+ User manual, Siemens (2020).

[21] ANSYS Fluent User's Guide, Release 17.2, ANSYS, Inc. (2016).

Problems

9.1. How are the basis functions selected for acoustics' finite element and BEMs?

9.2. Discuss various methods to calculate the undetermined coefficients in numerical approximation methods.

9.3. What is the purpose of shape function?

9.4. Describe the Finite Element approach for forced vibration analysis.

9.5. Discuss the advantages and disadvantages of mode superposition and direct analysis method in the FEM?

9.6. What type of nodal degree of freedom exists in the structural element, acoustic element?

9.7. Define Green's function.

9.8. What type of computational challenges need to be addressed in the BEM?

9.9. Describe the process map to predict the acoustic radiation using commercial tools.

9.10. What are the common element types used in the FEM and BEM meshing?

9.11. What type of mesh quality check must be done for the finite element analysis?

9.12. Describe the SEA process map.

9.13. What is the maximum element size to be maintained in the meshing for upper-frequency analysis of 3000 Hz and speed of sound 334 m/s?

[**Ans.: 18.56 mm**]

9.14. Calculate the modal density of the cavity with the dimensions of 0.610 m × 0.305 m × 6.1 m, speed of sound 340 m/s, and density 1.2 kg/m^3. The one-third octave band frequencies for calculation are 50 Hz and 5000 Hz.

[**Ans.: 0.002, 1.485**]

Nomenclature

Every symbol has been described at the place of its first appearance in the text. Therefore, only those symbols that appear often in the text, are described as follows.

a Acceleration

A Complex amplitude of the forward progressive wave; Area of cross-section; Total absorption of the room

AFR   Air fuel ratio

ANC   Active noise control

AVC   Active vibration control

B Number of cylinders of a compressor; Complex amplitude of the rearward progressive wave

BC   Boundary conditions

BEM   Boundary element method

BPF   Blade passing frequency

bw Half-power bandwidth

bw_n Bandwidth of the n-octave band

c Speed of sound; Damping coefficient

$[C]$ Damping matrix

c_c Critical damping

c_g Group velocity

c_p Phase velocity

CLD   Constrained layer damping

CTR   Concentric tube resonator

D	Daily noise dosage; Diffraction directivity factor; Muffler shell diameter
d	Diameter of the exhaust pipe or tail pipe
DAQ	Data acquisition system
DBEM	Direct boundary element method
dB	Decibel
dB(A)	A-weighted decibel
dBA	A-weighted decibel
DF	Directivity factor
DI	Directivity index
DNL	Day night average sound level
DOF	Degree of freedom
e	Eccentricity
E	Flow resistivity; Young's Modulus; Electromotive force; acoustical energy
ECTR	Extended concentric tube resonator
$E_{\max}$	Maximum element size
E_r	Storage modulus
ESM	Equivalent source method
E_t	Loss modulus
ETEC	Extended tube expansion chamber
F	Froude's friction factor; Force amplitude
f	Frequency (in Hertz); Force
$\{F\}$	Force vector
F_0	Amplitude of the exciting force
f_{dip}	Frequency at which a dip or trough occurs in the TL spectrum
F_f	Force transmitted to the compliant foundation
f_i	Lower limit of the frequency band
FEM	Finite element method
FIR	Finite impulse response
FLD	Free layer damping
f_m	Mean frequency of the band
$f_{m,n}$	Natural frequency corresponding to the (m, n) mode of vibration

f_p	Peak frequency
FR	Force ratio
F_T	Force transmitted to the foundation
f_u	Upper limit of the frequency band
FWH	Ffowcs Williams and Hawkings
G	Shear modulus; Green's function
g	Shear parameter
GE	Governing equation
G_2	Storage shear modulus of the viscoelastic layer in CLD treatment
H	Dynamic head; specific sound power of a fan
$[H]$	Dynamic matrix
h	Plate thickness; ratio of flow area to lined perimeter
I	Acoustic intensity; Second moment of area of cross-section (Moment of Inertia)
IBEM	Indirect boundary element method
IL	Intensity level; Insertion loss
I_{ref}	Reference intensity
K	Environmental correction; Dynamic pressure loss factor
KHIE	Kirchoff–Helmholtz integral equation
k	Wave number; Stiffness
$k_0 l$	Helmholtz number
K_a	Acoustical power coefficient
k_a	Axial stiffness
k_c	Convective wave number
k_s	Shear stiffness
k_x	Wave number in the x-direction
k_y	Wave number in the y-direction
k_z	Wave number in the z-direction
$[K]$	Stiffness matrix
l	Length
L_a	Acceleration level
L_d	Day-time average SPL; Displacement level
L_{dn}	Day-night average SPL
L_{eq}	Equivalent sound pressure level over a period

L_I	Intensity level
L_n	Night-time average SPL
l_p	Perforate length
L_p	Sound pressure level
L_{pA}	A-weighted sound pressure level; Sound level
L_u	Velocity level
L_w	Sound power level
L_x	x-percentage exceeded level
m	Area ratio; Mass (lumped)
M	Mean flow Mach number
$[M]$	Inertia matrix
n	Ratio of the area of cross-section of the narrower pipe to that of the wider pipe at a sudden area discontinuity; Speed order; modal density
$[N]$	Shape function matrix
NAH	Nearfield acoustic holography
N_{cyl}	Number of cylinders
N_i	Fresnel number in the ith direction
NR	Noise reduction
N_s	Strouhal number
n_s	Number of surfaces touching at the source
N_{st}	Number of strokes
$\{O\}$	Null vector
OAF	Open area fraction
OAR	Open area ratio
P	Power of the motor driving a pump; Lined perimeter; Loudness level in phons
PML	Perfect matched layer
p	Acoustic pressure
p_0	Atmospheric pressure
p_s	Source pressure
p_{th}	Threshold pressure
Q	Directivity factor; source strength
Q_B	Barrier directivity factor

$q_i(t)$	Modal coordinate for the ith mode
Q_l	Locational directivity factor
R	Resistance; Real part of the impedance; Gas constant; Reflection coefficient; Non-dimensional acoustic flow resistance; Room constant
r	Radial coordinate; Radial distance; Frequency ratio
R_2	Room constant of the receiver room
RPM	Revolutions per minute
S	Area of cross-section; Loudness index in sones; Free-flow area of the cross-section; microphone sensitivity
SEC	Simple expansion chamber
SEL	Sound exposure level
S_m	Area of the hypothetical measurement surface; microphone sensitivity
SPL	Sound pressure level
SSL	Source strength level
STC	Sound transmission class
S_w	Surface area of the partition wall
SWL	Sound power level
T	Temperature; Time period
t	Time variable
$[T]$	Transfer matrix
T_{ij}	The ith row jth column element of the transfer matrix; Lighthill stress tensor
$T_{n,m}$	Transfer factor
T_{60}	Reverberation time
TFM	Transfer function method
TL	Transmission loss
TL_a	Axial TL
TL_{net}	Net TL
TL_{tp}	Transverse power TL
TR	Transmissibility
U	Mean flow axial velocity
u	Particle velocity

$\{u_i\}$ Modal vector for the ith mode

u_0 Initial velocity

v Volume velocity; Mass velocity; Velocity of a lumped mass

V Volume; complex amplitude of velocity

W Power flux

w Transverse (flexural) displacement

W_{ref} Reference power

x Open area ratio; Displacement (instantaneous)

X Reactance; Imaginary part of the impedance; Displacement amplitude

x_0 Initial displacement

x_{st} Static displacement or deflection

$\dot{x}$ Instantaneous velocity

$\{\ddot{x}\}$ Acceleration vector

Y Characteristic impedance; Amplitude of the displacement excitation

$y(t)$ Support displacement

z Axial coordinate

Z Impedance

Z_c Lumped impedance of the cavity; Characteristic impedance

Z_e Lumped impedance of the exhaust pipe

Z_L Load impedance

Z_s Source impedance

Z_t Lumped Impedance of the tail pipe

$\ominus$ Logarithmic subtraction

$\oplus$ Logarithmic addition

Subscripts

a	Absorber; Axial; Acoustic
b	Bending
d	Downstream
e	Effective
eq	Equivalent
ext	Exterior
f	Flexural
g	Geometric or physical
i	Incident; The ith mode
int	Interior
m	Mechanical
n	Of the neck; Natural; normal
r	Reflected
ref	Reference value
rms	Root mean square
s	Sample
t	Transmitted
u	Upstream

Greek Symbols

α	Absorption coefficient of a partition wall
$\overline{\alpha}$	Absorption coefficient of an acoustic layer
δ	End correction; Logarithmic decrement; Dirac delta function
δ_i	Difference in the diffracted path and the direct path in the ith direction
Δ	Differential length
Δp	Stagnation pressure drop; Back pressure
γ	Ratio of specific heats
λ	Wave length
υ	Poisson's ratio
Ω	Angular speed of rotation; domain
ω	Circular frequency (in radians/second)
ω_d	Damped natural frequency

ω_n Natural frequency (undamped)

ρ_0 Atmospheric density

σ Porosity of the perforate

τ Transmission coefficient

ξ Displacement

η_{G_2} Loss factor of the viscoelastic material in shear in CLD treatment

η Loss factor; non-dimensional frequency parameter

η_{ij} Coupling loss factor

ζ Acoustic impedance of a passive subsystem at the upstream end; Damping ratio

Index

sheet metal components, 126, 161, 306
shell cavity, 317
simple expansion chamber, 245, 250–252, 255–257, 259, 272, 279, 280, 282, 297
solver, 386
Sommerfeld radiation condition, 359
sonicjet, 338
sound absorber, 310
sound barrier, 310
sound exposure level, 67
sound field in a room, 200
sound intensity level, 12
sound intensity method, 71
sound intensity probe, 77
sound level meter, 5, 19, 27, 58, 203
sound power level, 12, 21, 29, 200, 204, 222, 270, 312–316, 319, 321–324, 326–327, 330–334, 343, 346
sound pressure level, 12, 15–16, 18–19, 21, 28, 199–200, 222, 270, 278, 312, 315–316, 321, 324, 331–332, 346
average, 21, 203
sound pressure method, 72
sound speed, 1, 3, 119, 153, 203, 273
sound transmission class, 182
Sound-Hard Boundary, 357
Sound-Soft Boundary, 357
source characteristics, 272–274, 278
spacer size, 77
specific flow resistance, 184
specific fuel consumption (SFC), 295
specific power levels, 318
specific sound power, 317–319
specific transmission loss, 266–267
speech range, 13
SPL measurement, 12, 311
spur gears, 307, 341
stagnation pressure drop, 261, 271, 294–295, 297, 310, 318
stand-alone acoustic enclosure, 310, 320–321

standing wave analysis, 293
state variables, 1, 237
static pressure rise, 318, 320
Statistical energy analysis, 353, 373
steam turbine, 326
stepped dies, 307
stiffeners, 160
stiffness controlled
region, 209
TL, 287
stiffness matrix, 105
stillage, 308, 344, 346
storage modulus, 132, 162–163
Strouhal number, 284, 337
structural discontinuities, 310
structural vibration, 289, 293, 324
structure-borne sound, 310, 325
subsystem, 374–377, 379–381
suction muffler, 317
sudden contraction, 249, 252, 254, 296–297
sudden expansion, 238, 249, 252, 254, 263, 296
superheated steam, 284, 309
surface impedance, 189
survey method, 29, 73, 75, 316

T

tail pipe, 3, 235–236, 252, 279–283, 286, 297, 346
temperature controlled drive, 320
termination, 3, 116, 233
anechoic, 3, 286
passive, 3
rigid, 3
thermal characteristic length, 184
thermal drive, 320
thermal foam, 264
Thevenin theorem, 234, 272
threshold of hearing, 1
time period, 2, 19, 98, 100
time weighting, 59
time-harmonic response, 365

Printed in the United States
by Baker & Taylor Publisher Services